D0216230

PROBLEM LAYOUTS FOR TECHNICAL DESCRIPTIVE GEOMETRY

SECOND EDITION

B. LEIGHTON WELLMAN

Professor of Mechanical Engineering and
Head of the Division of Engineering Drawing
Worcester Polytechnic Institute

789101112—GK—10987

McGRAW-HILL BOOK COMPANY, INC.

New York · Toronto · London · 1957

INSTRUCTIONS

These problem layouts have been specially prepared for use with the second edition of "Technical Descriptive Geometry" by B. Leighton Wellman (McGraw-Hill Book Company, Inc., 1957). The problems have been selected from those in the text, and each layout accurately duplicates the conditions stated in the correspondingly numbered text problem. Thus the 111 layout sheets of the set provide an economical and timesaving device for covering all the topics that are essential to a complete and well-balanced course in descriptive geometry. Blank sheets have been included for special problems that the instructor may wish to assign.

Each sheet is numbered in the lower right corner, and this number corresponds with the similarly numbered problem in the text. Sheet 7•9, for example, is the layout for the ninth problem in Group 7 in the problem chapter of the text (see page 412 of the text). Before starting the problem solution, always read the general statement for that group of problems, then the statement for the specific problem. Be sure also that you have read and studied the text articles listed for that group. Sheets marked with a group number and a dash (2•—, for example) indicate that there are two or more problems from that group on the same sheet. Sheets marked only with a dash indicate that there are two or more problems from different groups on the same sheet. In either case each problem is individually numbered.

In order to obtain accurate solutions the sheets must be carefully aligned with the T square when they are attached to the board. To do this match the edge of the T square blade with the short line-up marks that are placed on the inside of the borders (see Sheets 3•1 and 3•9). On sheets containing two or more problems, align the T square with the line separating the spaces (see Sheet 9•—). Do not use the borders to align the sheets. Although some problems can be solved in several different ways, the position of the given views on the sheet will usually indicate the desired form of solution. When the solution has been completed, scale and record the required distances or angles, properly dimensioning each item to show where on the drawing the answer was obtained (see page 414 of the text).

In the title strip of each sheet the draftsman should neatly letter the date, his name, and in the small space following the name, his seat, file, or division number as the instructor may require. The name of the school or department may be lettered in the long upper space. Guide lines for lettering (1/8-inch high) should be drawn through the small dash marks that are located near the center of the title strip. The instructor may use the grading scale at the left end of the strip by punching or marking out the proper numbers (marking 8 and 1/2 indicates a grade of 85 per cent).

CONTENTS

LINE	TYPE
AB	
AK	
KJ	
AJ	
BC	
BM	
MH	
MN	
HG	
CN	
NG	
GF	

1•16

SURFACE	TYPE
ABMKJHG	
BCDNM	
KMN	
JKNL	
DELN	
JHL	
EFGHL	

1•21

TOP	FRONT	SIDE
A		
B		
C		
	D	
	E	
	F	
		G
		H
		J

1•25

2	3	4	5	6		MULTIVIEW DRAWINGS	1•—
7	8	9	10	½	Scale / Date	Name	

2•1
2•8
2 3 4 5 6
7 8 9 10 ½ Scale Reduced
Date
PRINCIPAL VIEWS
Name
2•—

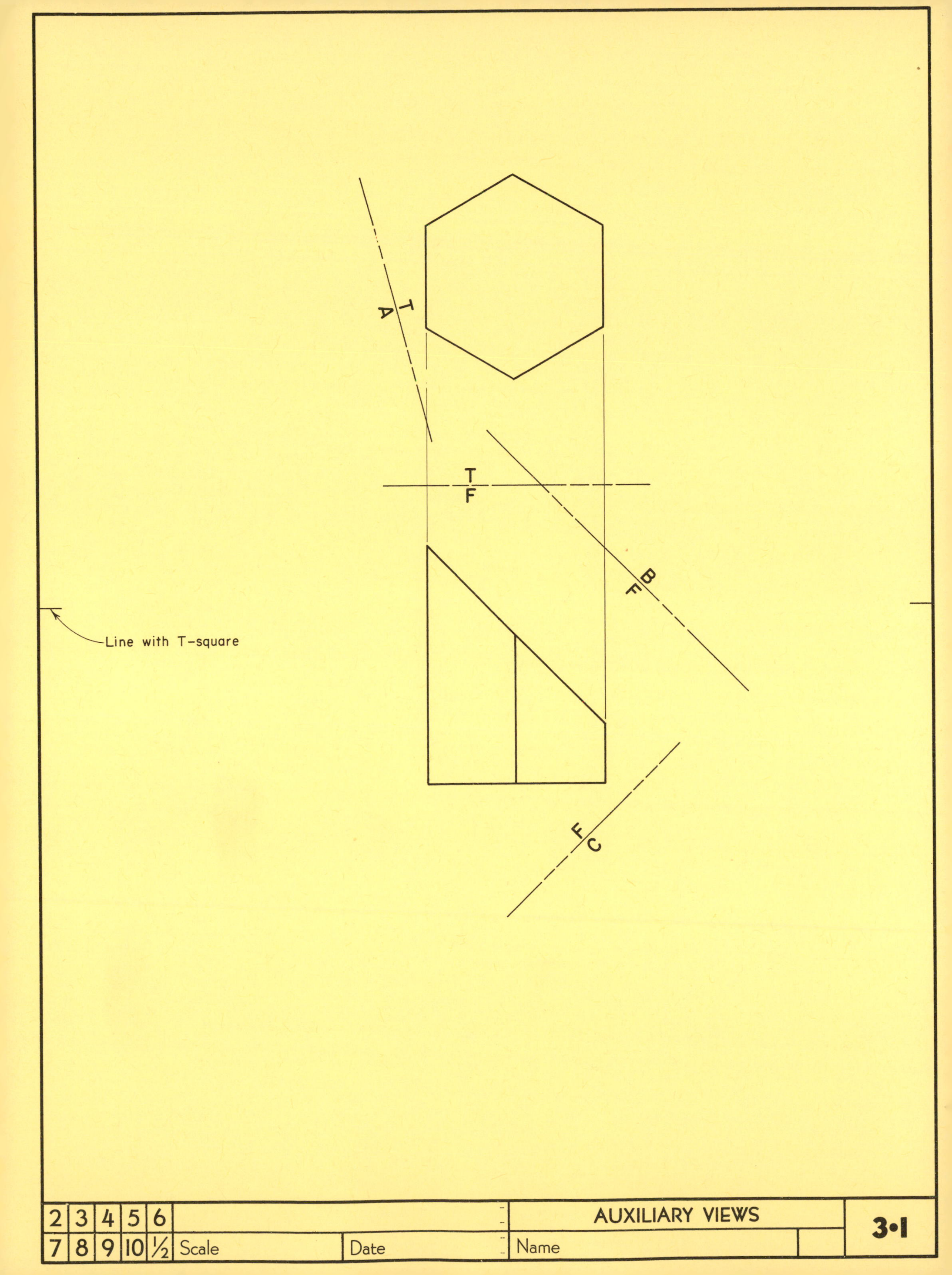
T
A
T
F
B
F
F
C
Line with T-square
2 3 4 5 6
7 8 9 10 ½
Scale
Date
AUXILIARY VIEWS
Name
3•1

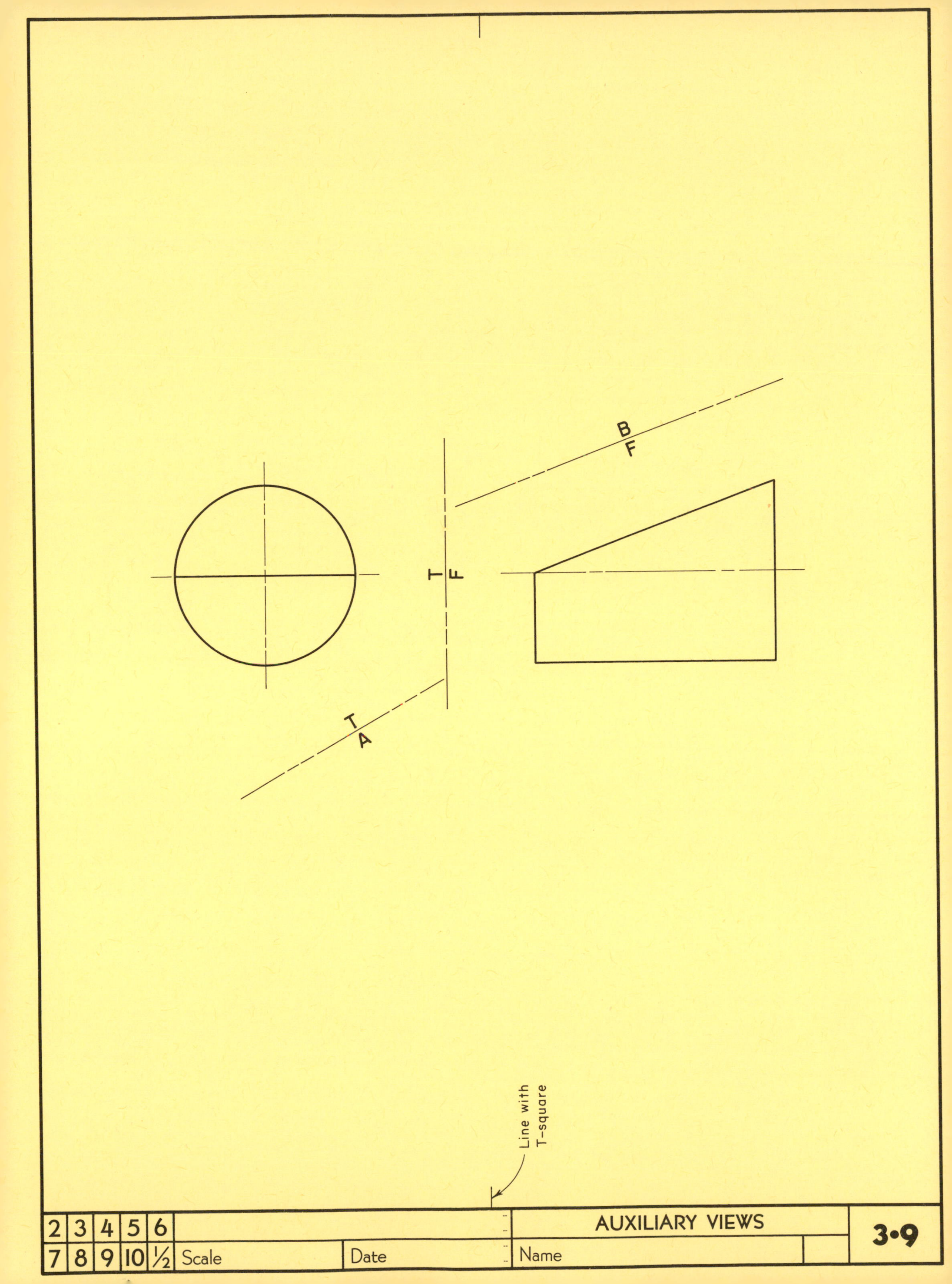
B
F
T
F
T
A
Line with
T-square
2 3 4 5 6
7 8 9 10 ½
Scale
Date
Name
AUXILIARY VIEWS
3•9

T
A

T
B

T
F

F
L

F
C

2	3	4	5	6		AUXILIARY VIEWS	3•11
7	8	9	10	½	Scale	Date	Name	

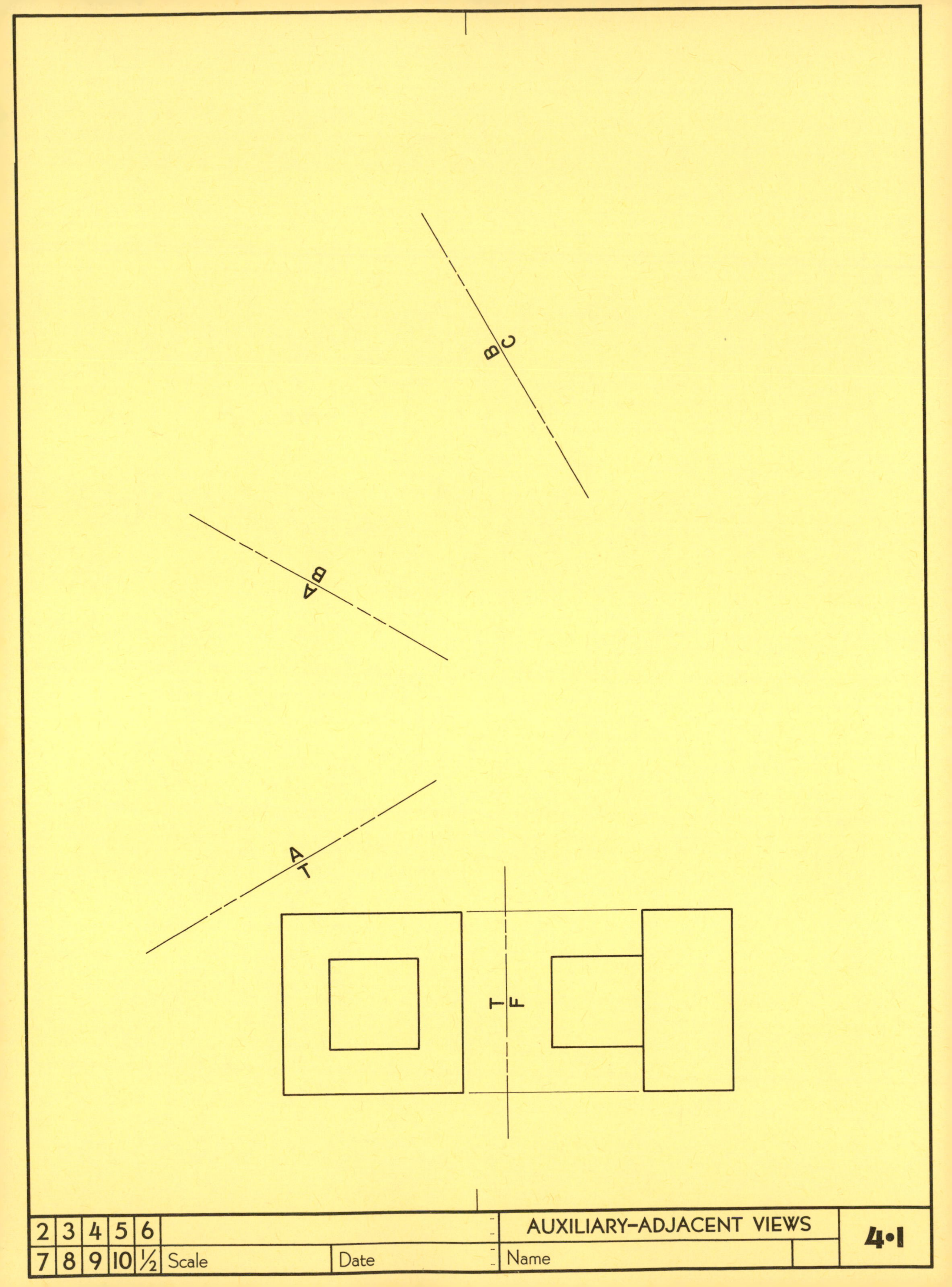
B
C
A
B
A
T
T
F
2 3 4 5 6
7 8 9 10 ½
Scale
Date
AUXILIARY-ADJACENT VIEWS
Name
4·1

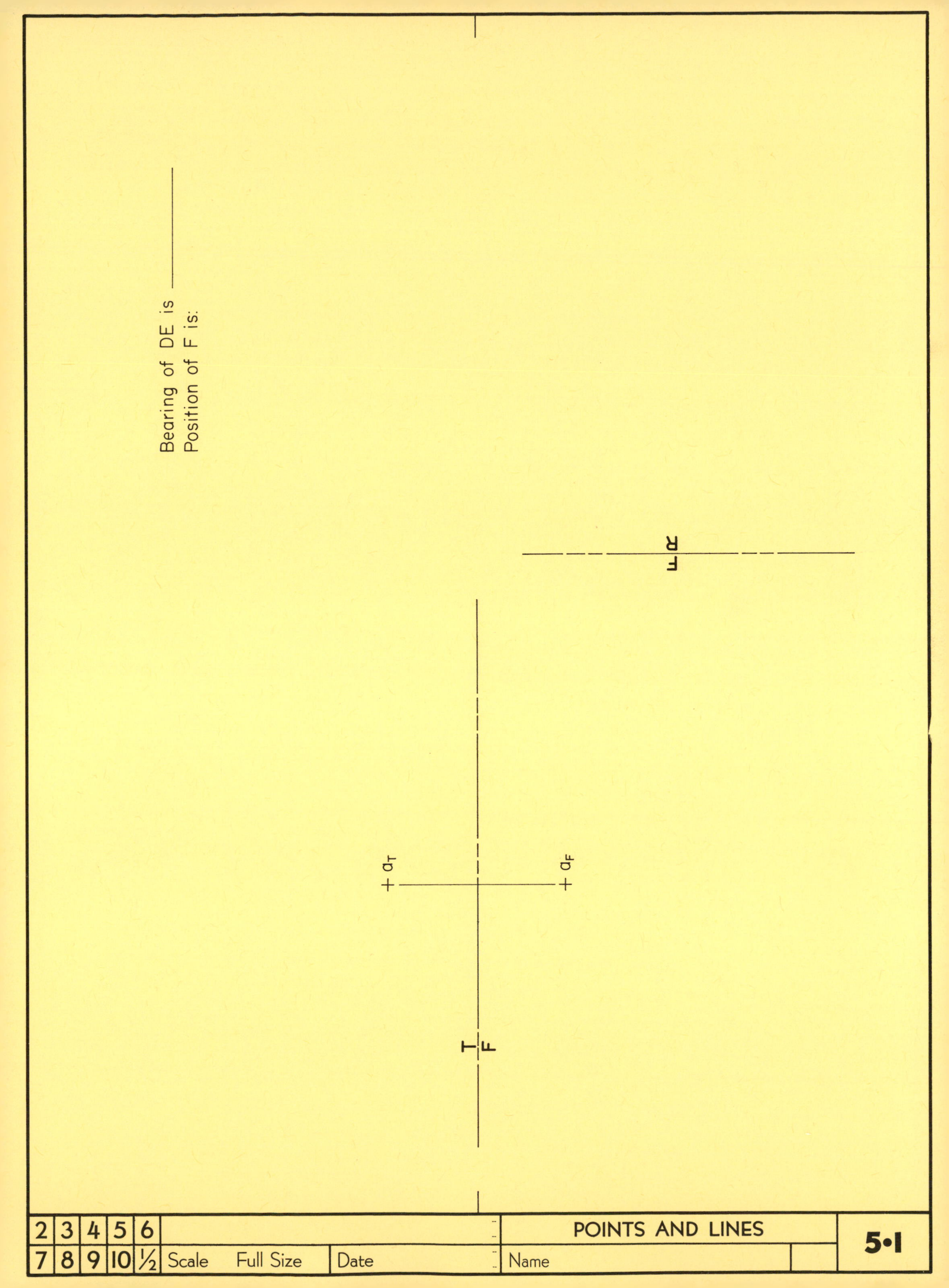
Bearing of DE is
Position of F is:
R
F
a_T
a_F
T
F
2 3 4 5 6
7 8 9 10 ½
Scale Full Size
Date
POINTS AND LINES
Name
5•1

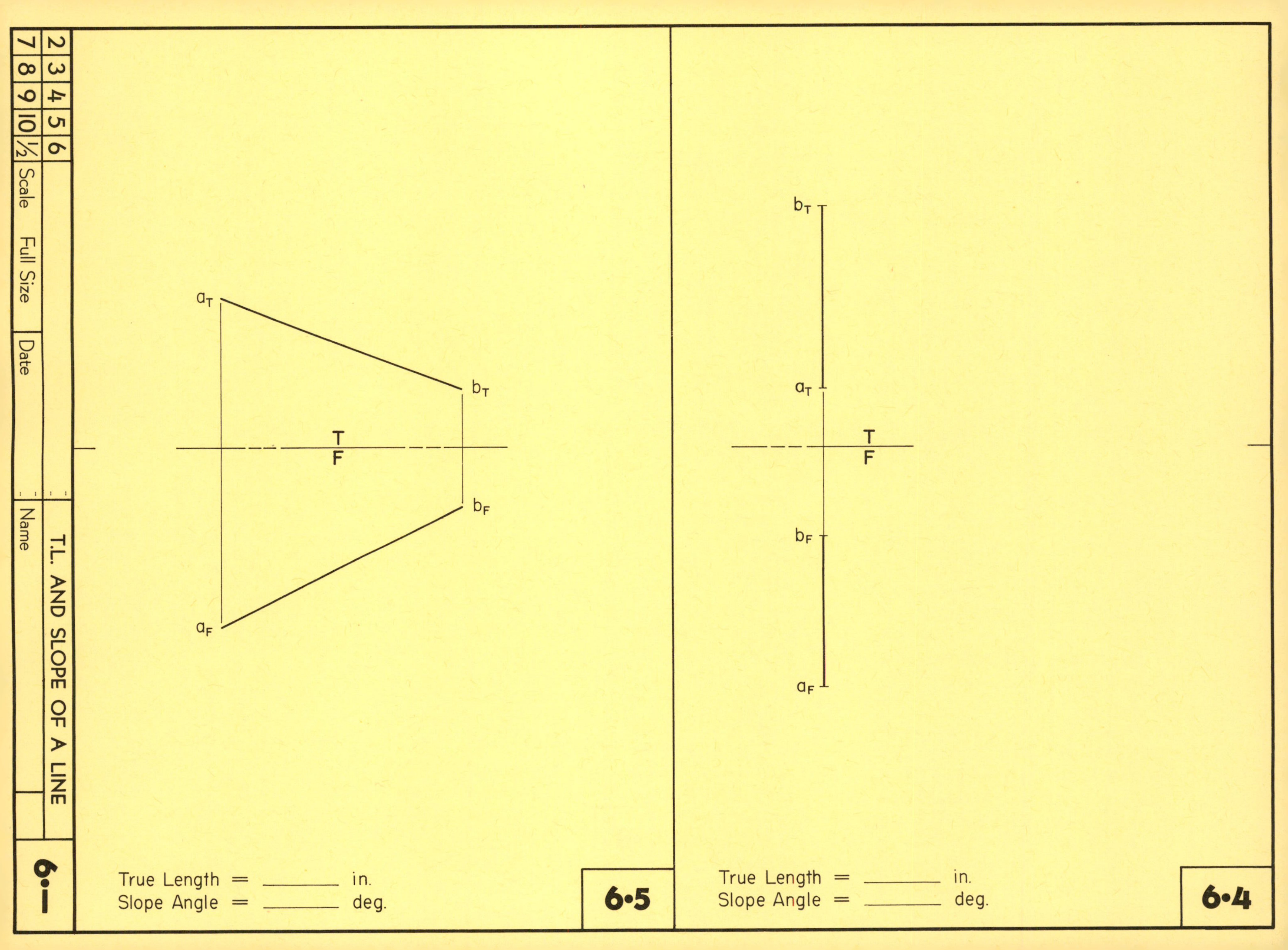
2 3 4 5 6
7 8 9 10 ½
Scale Full Size
Date
Name
T.L. AND SLOPE OF A LINE
6•—
a_T
b_T
T
F
b_F
a_F
True Length = ______ in.
Slope Angle = ______ deg.
6•5
b_T
a_T
T
F
b_F
a_F
True Length = ______ in.
Slope Angle = ______ deg.
6•4

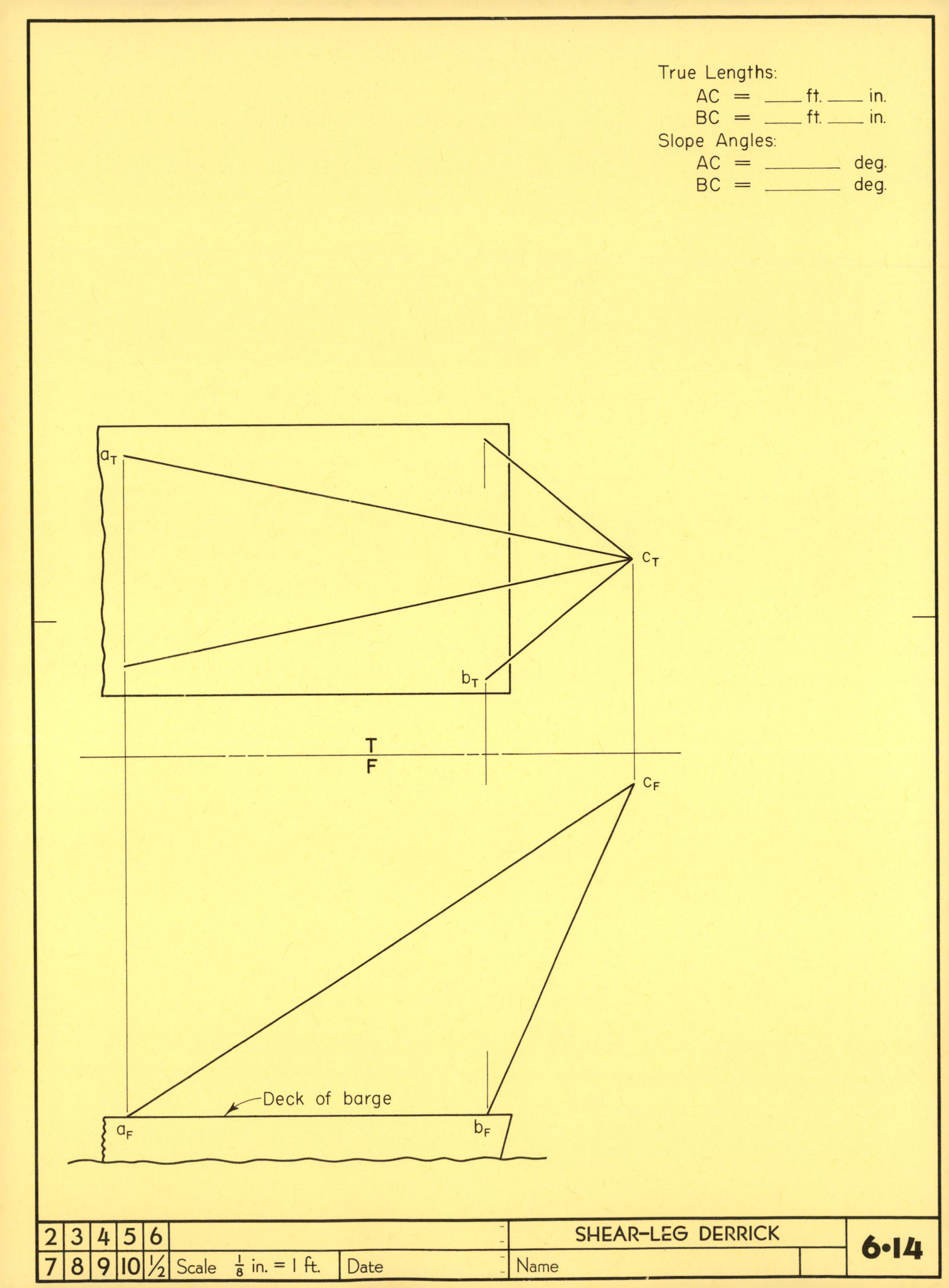
True Lengths:
AC = ____ ft. ____ in.
BC = ____ ft. ____ in.
Slope Angles:
AC = ________ deg.
BC = ________ deg.
a_T
c_T
b_T
T
F
c_F
Deck of barge
a_F
b_F
2 3 4 5 6
7 8 9 10 ½
Scale 1/8 in. = 1 ft.
Date
Name
SHEAR-LEG DERRICK
6•14

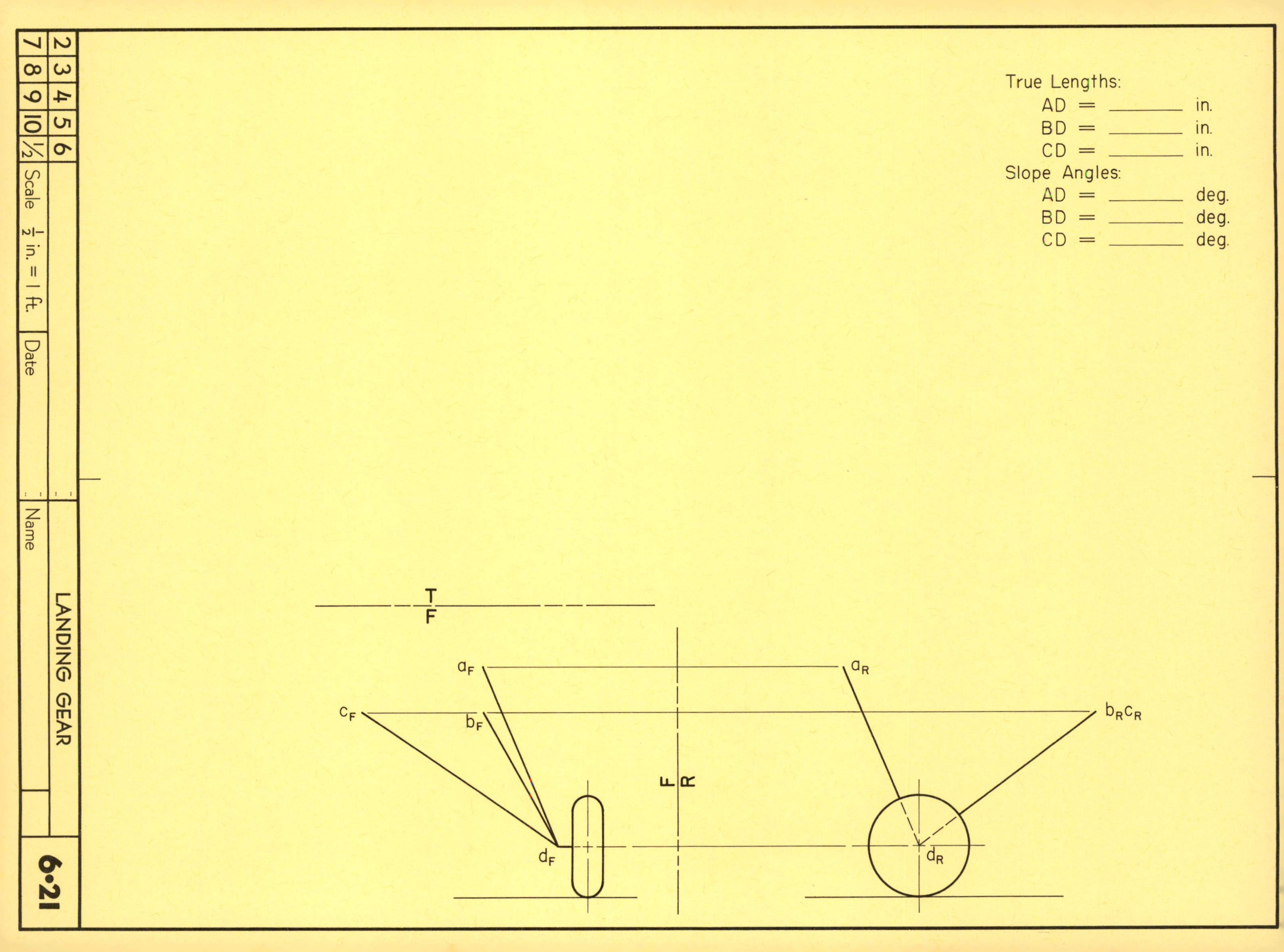

2	3	4	5	6		LANDING GEAR	6•21
7	8	9	10	½	Scale ½ in. = 1 ft. Date	Name	

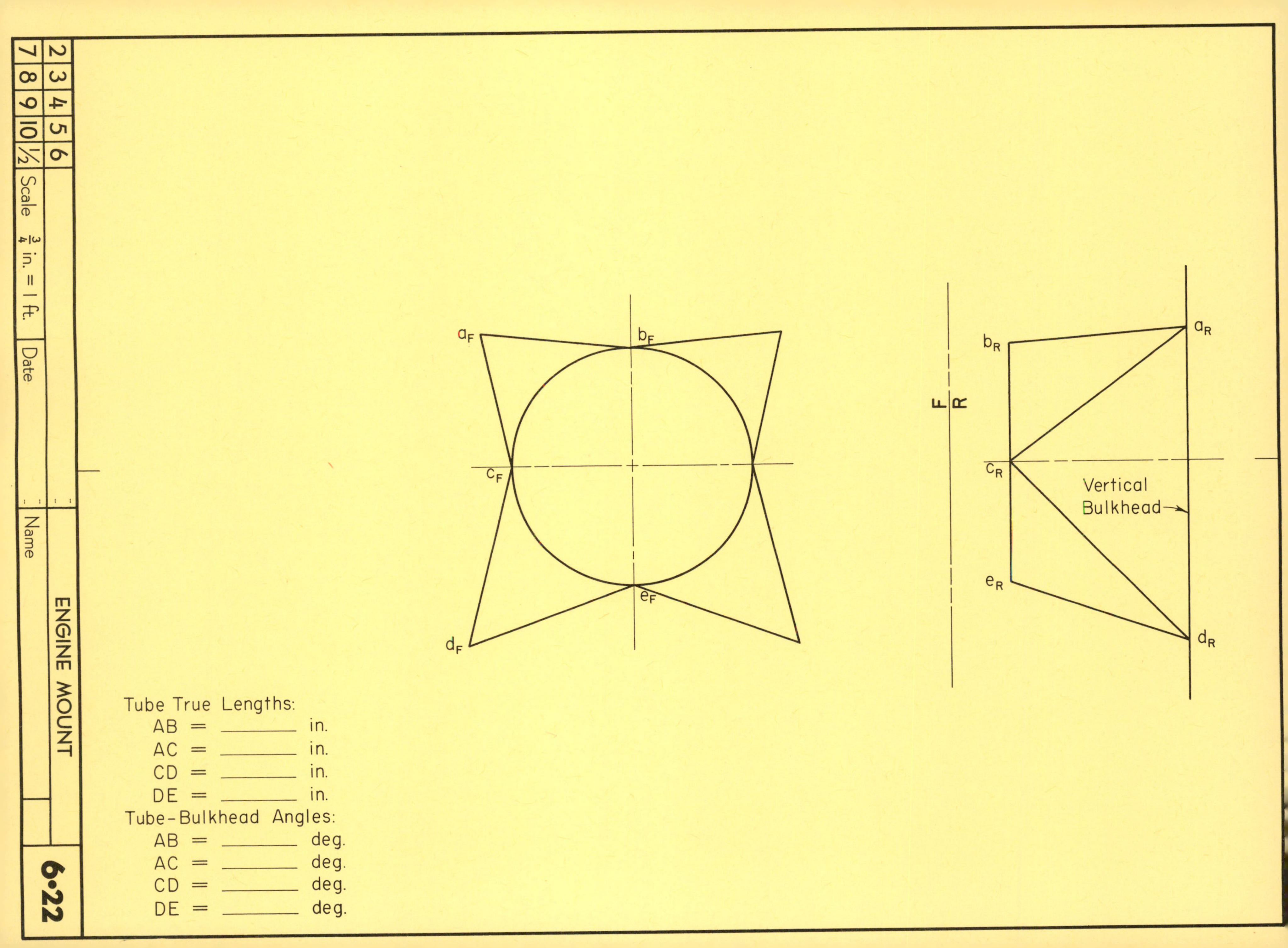

a_F
b_F
c_F
d_F
e_F
F
R
a_R
b_R
c_R
d_R
e_R
Vertical Bulkhead
Tube True Lengths:
AB = ______ in.
AC = ______ in.
CD = ______ in.
DE = ______ in.
Tube-Bulkhead Angles:
AB = ______ deg.
AC = ______ deg.
CD = ______ deg.
DE = ______ deg.
2 3 4 5 6
7 8 9 10 ½
Scale $\frac{3}{4}$ in. = 1 ft.
Date
Name
ENGINE MOUNT
6•22

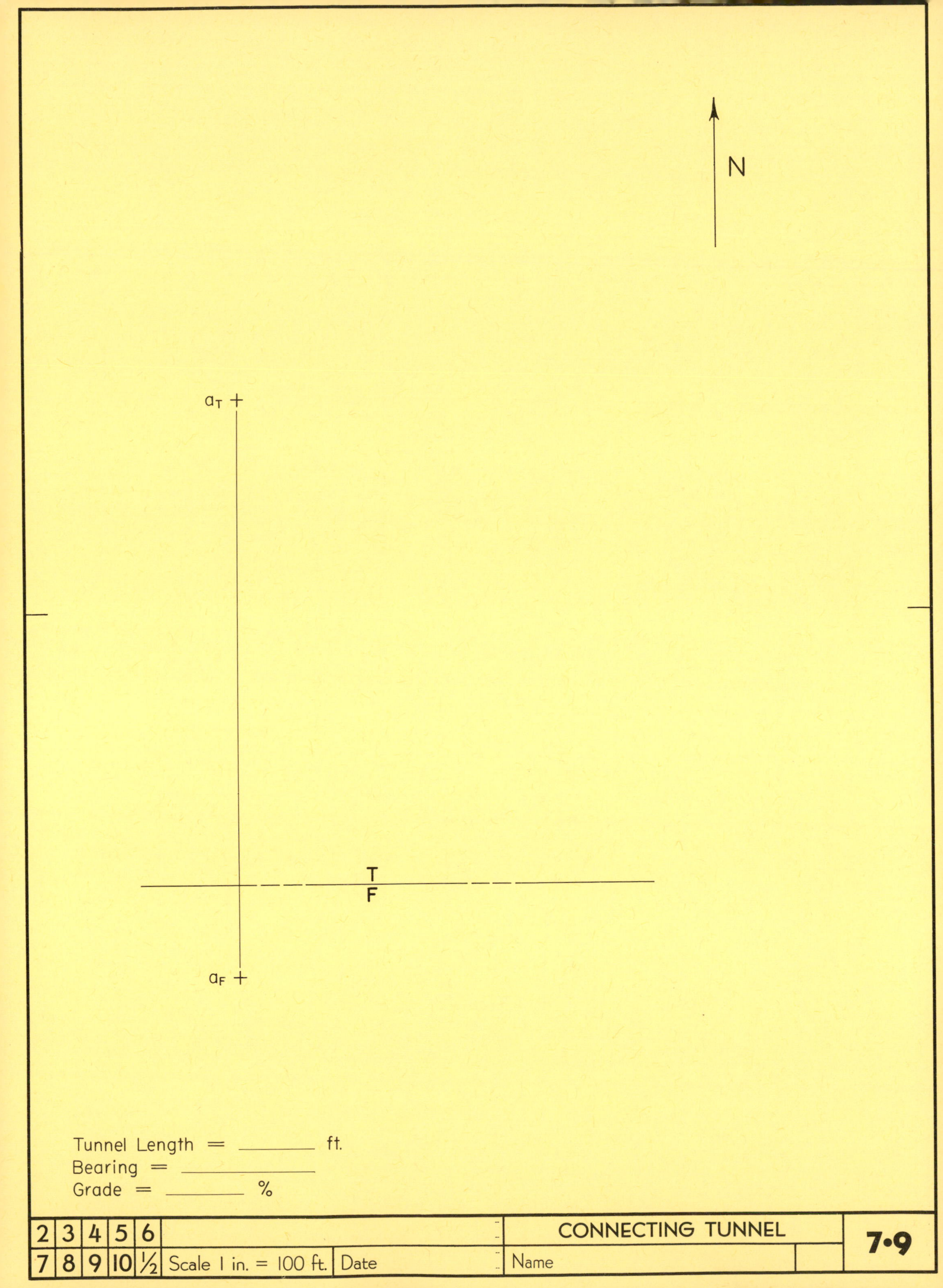
N
a_T
T
F
a_F
Tunnel Length = ________ ft.
Bearing = ____________
Grade = ________ %
2 3 4 5 6
7 8 9 10 ½
Scale 1 in. = 100 ft.
Date
CONNECTING TUNNEL
Name
7•9

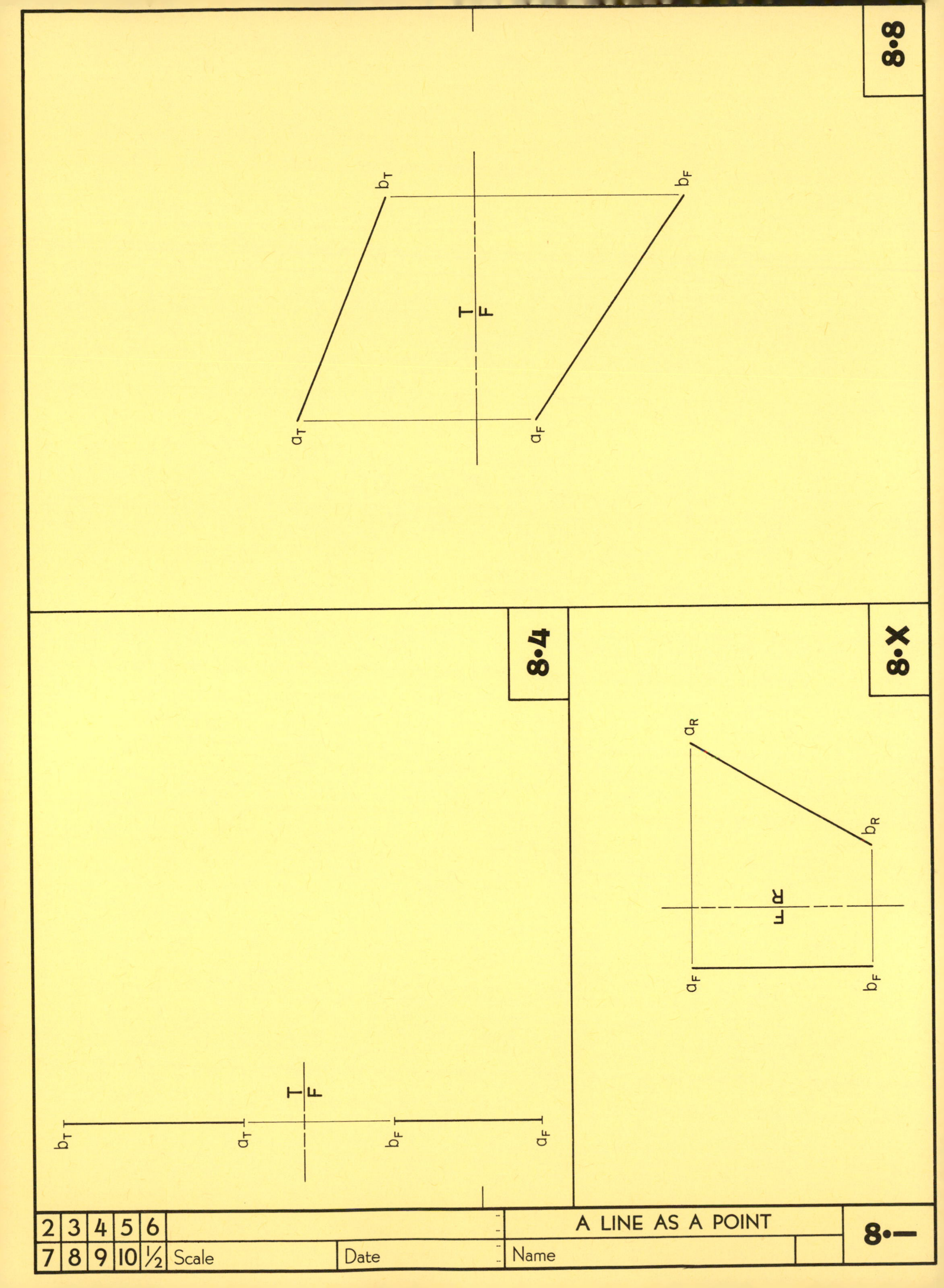
8·8
T
F
a_T
b_T
a_F
b_F
8·4
T
F
b_T
a_T
b_F
a_F
8·X
a_R
b_R
R
F
a_F
b_F
2 3 4 5 6
7 8 9 10 ½
Scale
Date
A LINE AS A POINT
Name
8•—

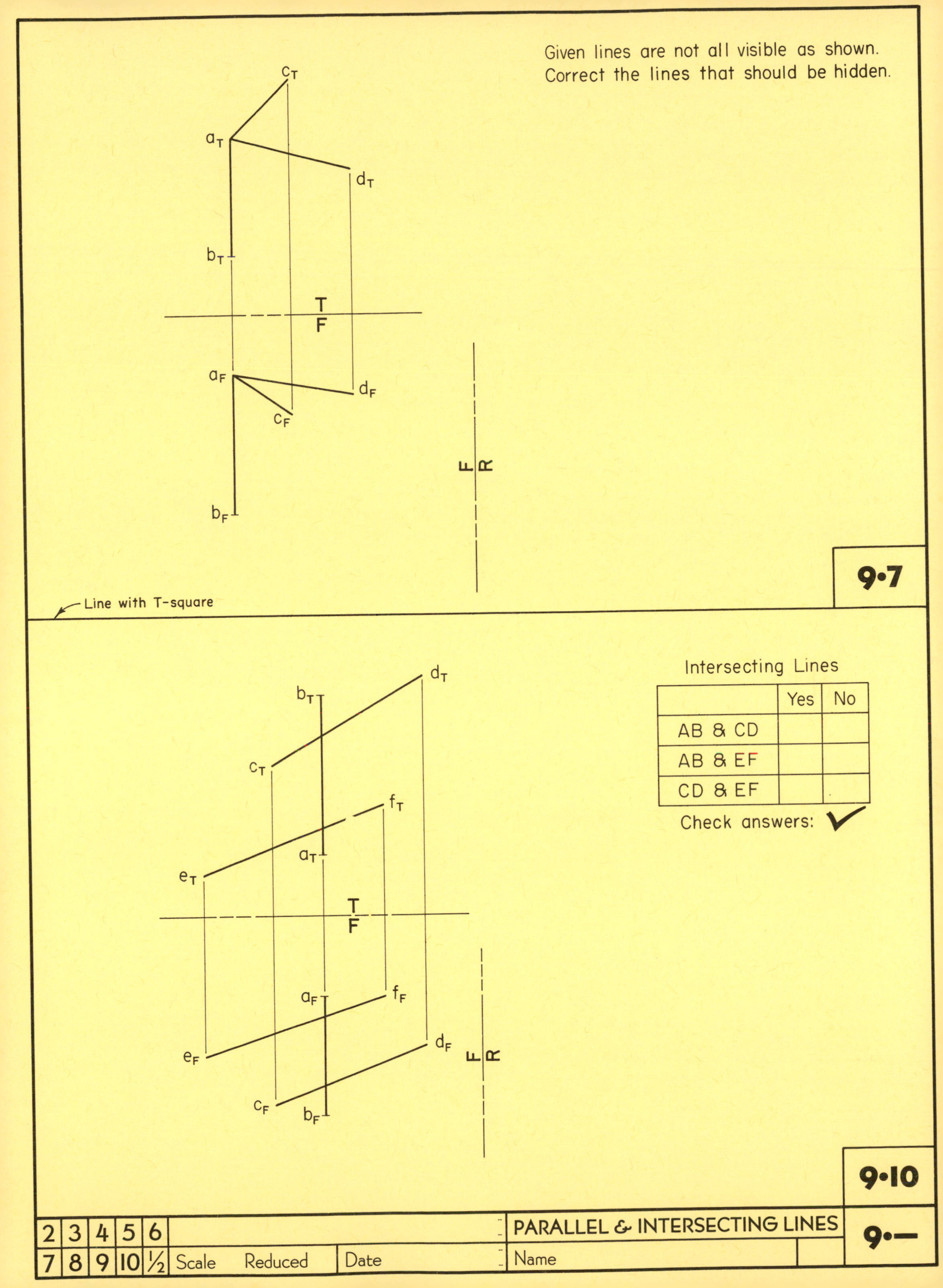

Intersecting Lines

	Yes	No
AB & CD		
AB & EF		
CD & EF		

Check answers: ✓

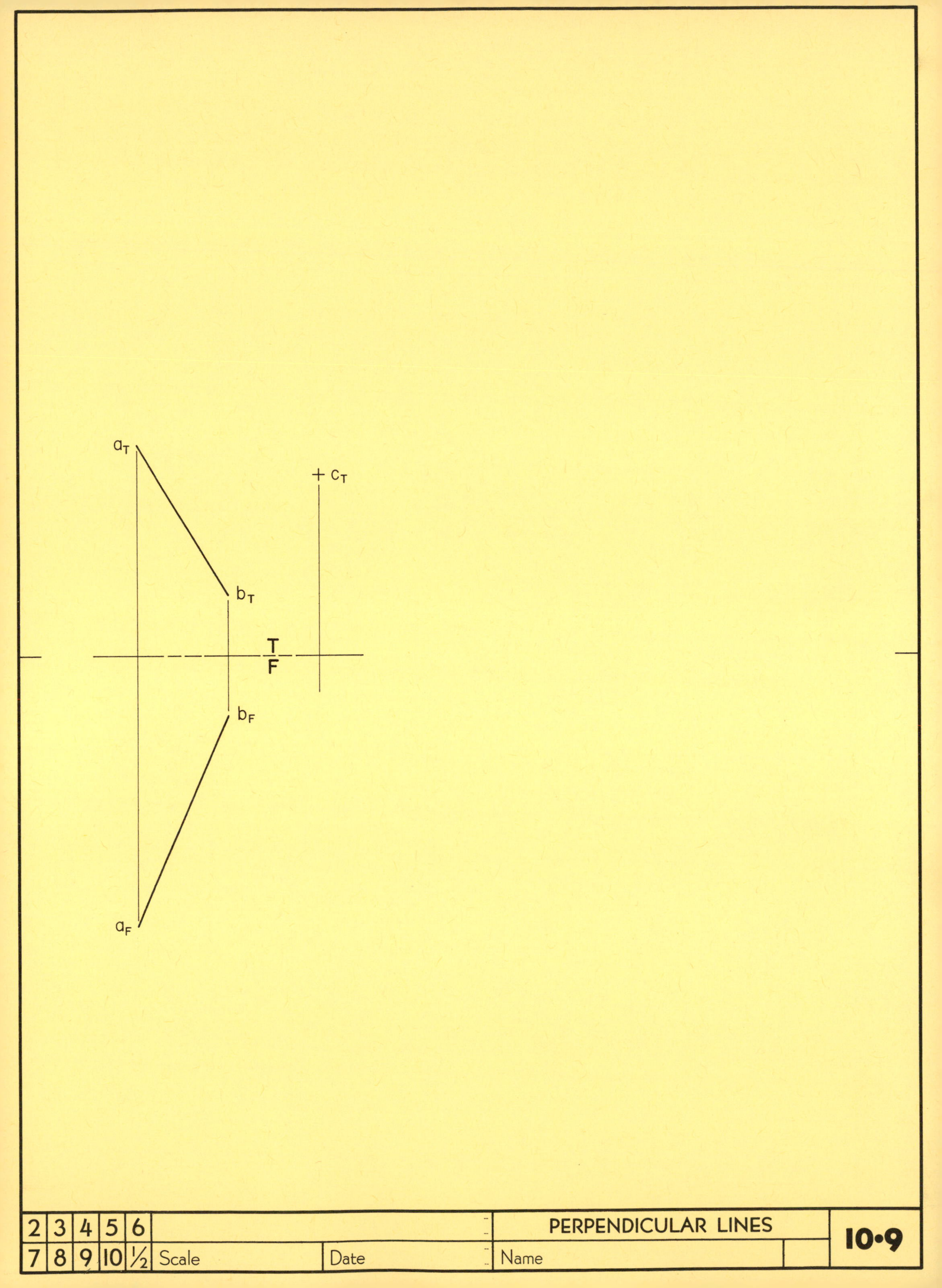

aT
+ cT
bT
T
F
bF
aF
2 3 4 5 6
7 8 9 10 ½
Scale
Date
PERPENDICULAR LINES
Name
10•9

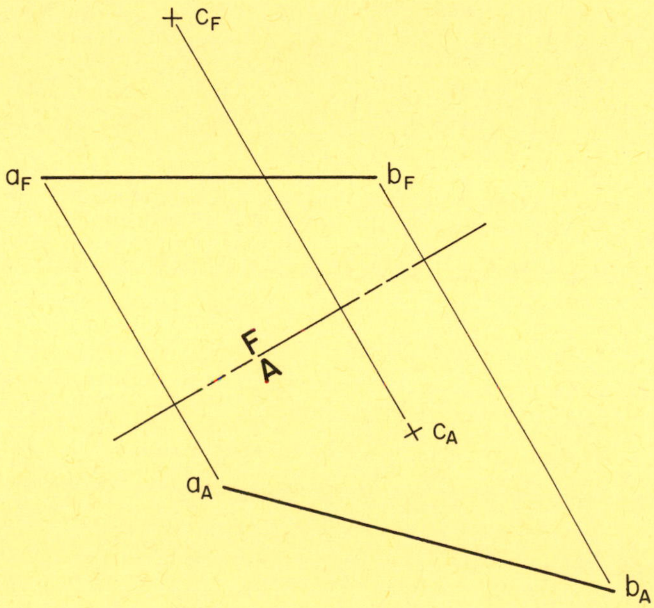

c_F

a_F

b_F

F
A

c_A

a_A

b_A

True Length = ________ in.
Bearing = ________
Slope Angle = ________ deg.

2	3	4	5	6		DISTANCE – POINT TO LINE	11•16
7	8	9	10	½	Scale Full Size Date	Name	

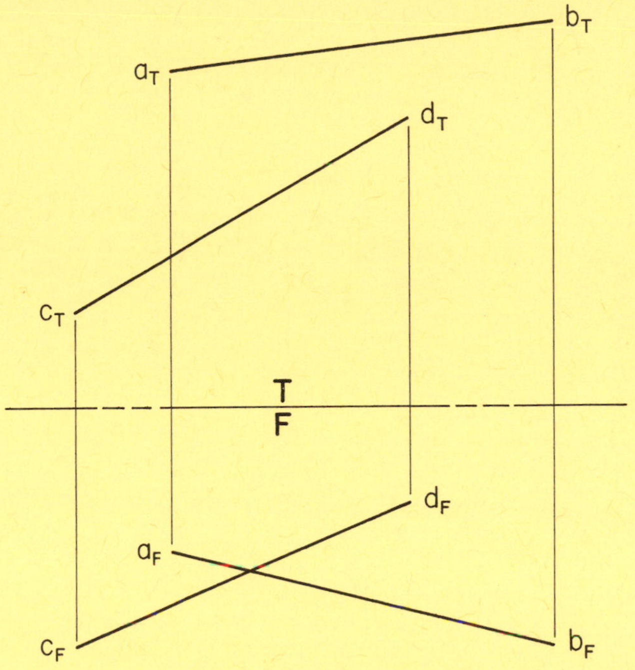

b_T

a_T

d_T

c_T

T

F

d_F

a_F

c_F

b_F

True Length = ________ in.

Bearing = ____________

Slope Angle = ________ deg.

2	3	4	5	6		DISTANCE BETWEEN SKEW LINES	12•5
7	8	9	10	½	Scale Full Size / Date	Name	

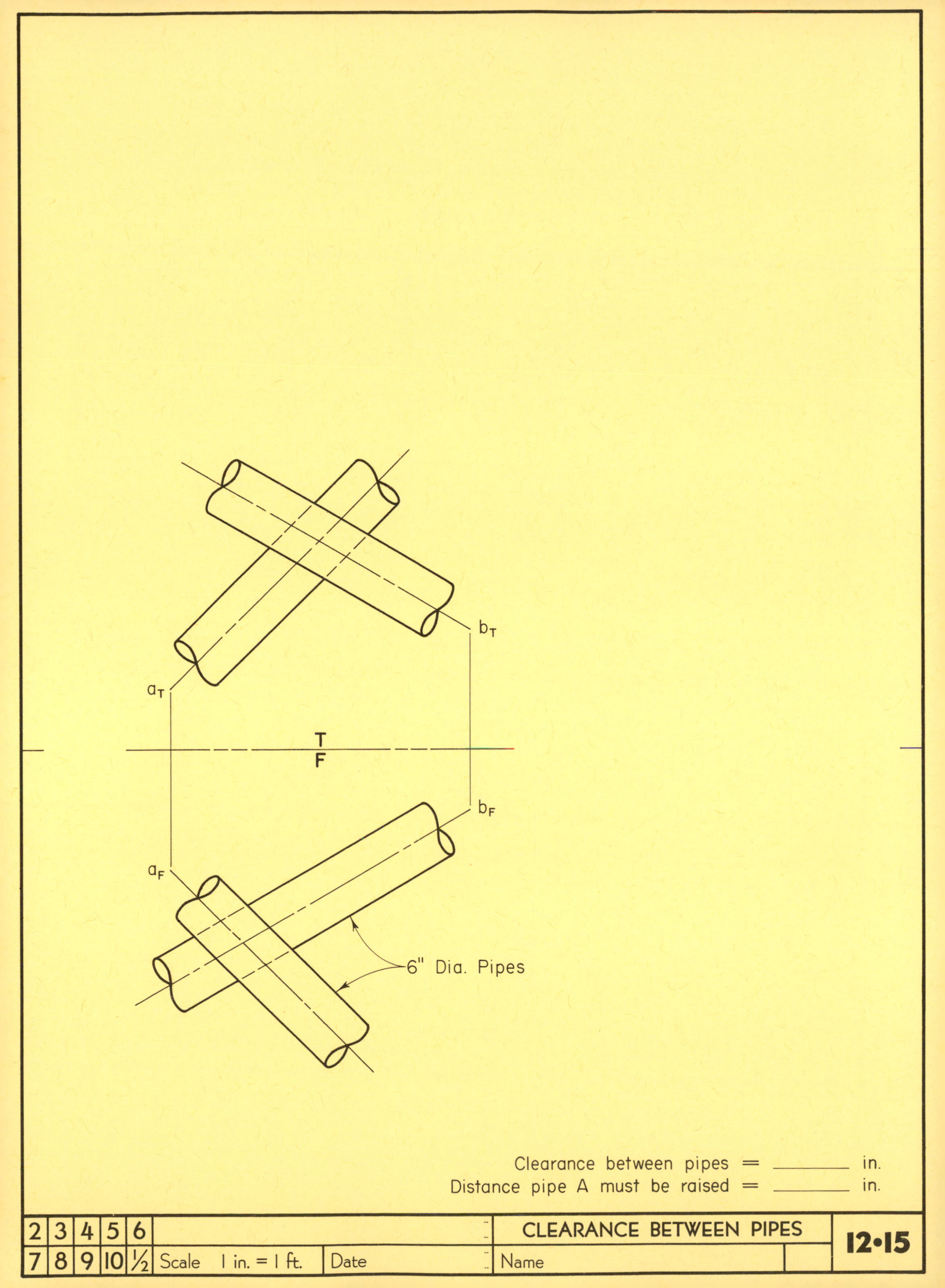

b_T
a_T
T
F
b_F
a_F
6" Dia. Pipes
Clearance between pipes = ________ in.
Distance pipe A must be raised = ________ in.
2 3 4 5 6
7 8 9 10 ½
Scale 1 in. = 1 ft.
Date
CLEARANCE BETWEEN PIPES
Name
12•15

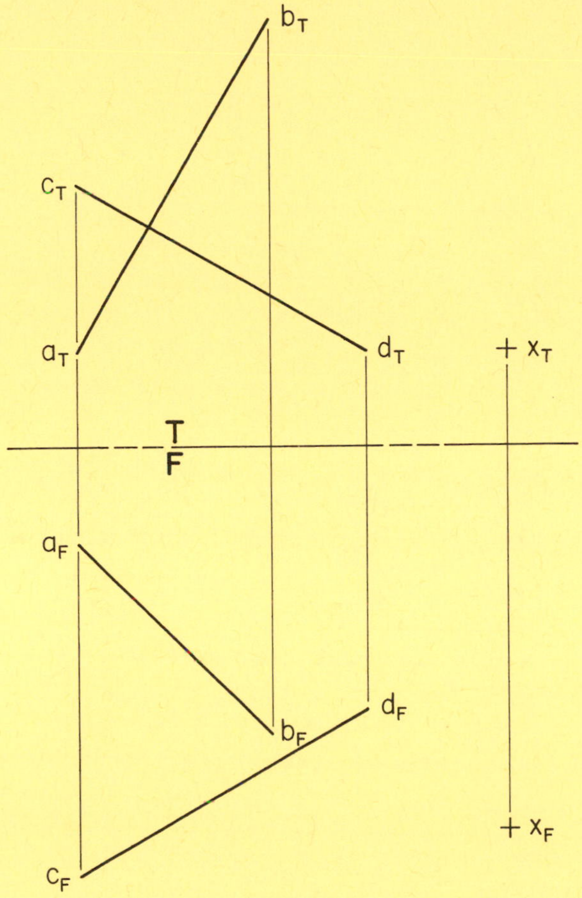

b_T

c_T

a_T

d_T

$+ x_T$

T

F

a_F

d_F

b_F

$+ x_F$

c_F

2	3	4	5	6			LINE THRU POINT & SKEW LINES	13·6
7	8	9	10	½	Scale	Date	Name	

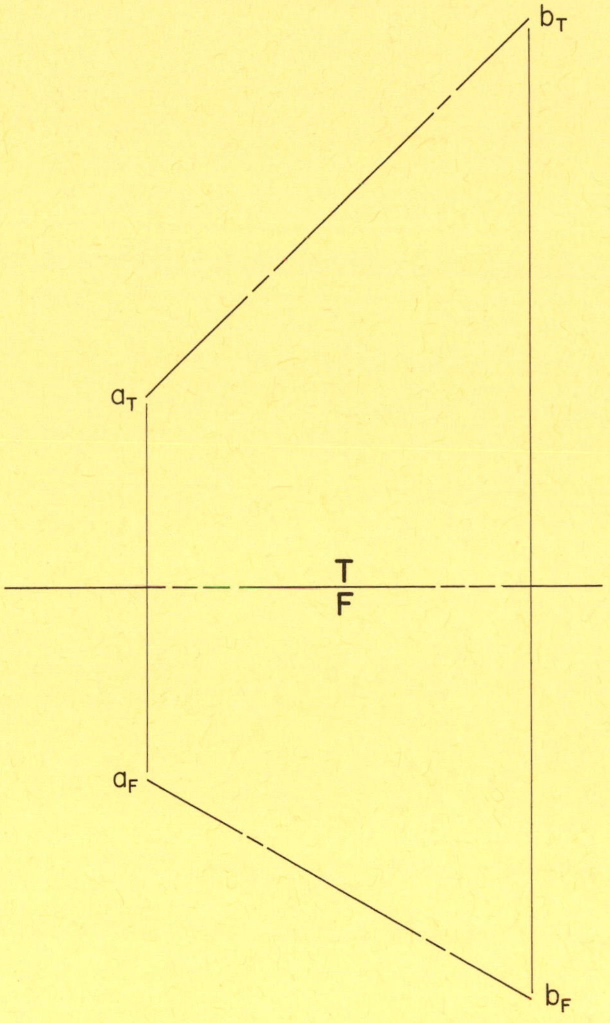

b_T

a_T

T

F

a_F

b_F

2	3	4	5	6		PYRAMID ON INCLINED AXIS	14·3
7	8	9	10	½	Scale Full Size Date	Name	

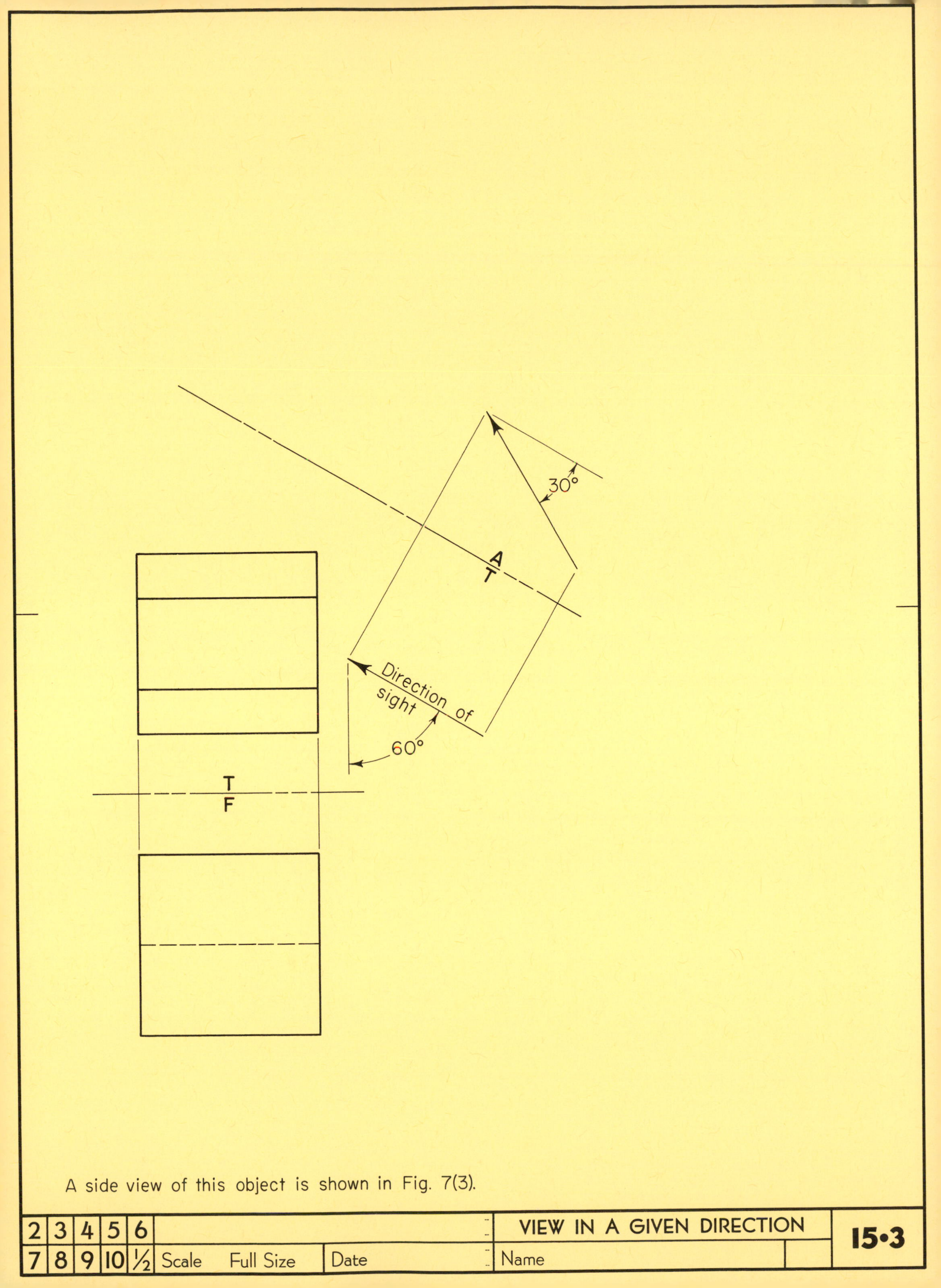
30°
A
T
Direction of
sight
60°
T
F
A side view of this object is shown in Fig. 7(3).
2 3 4 5 6
7 8 9 10 ½
Scale Full Size
Date
VIEW IN A GIVEN DIRECTION
Name
15•3

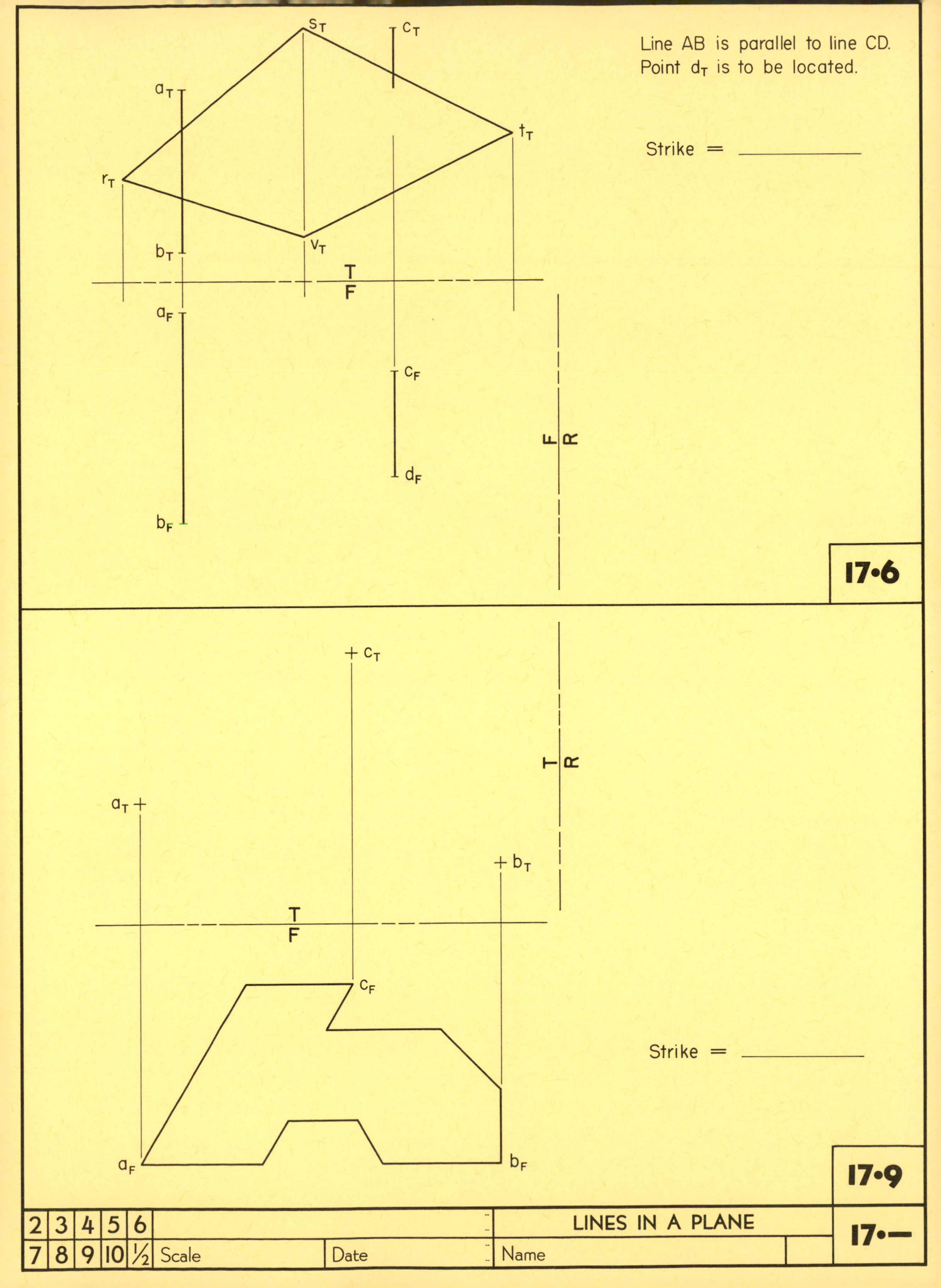

s_T
c_T
Line AB is parallel to line CD.
Point d_T is to be located.
a_T
t_T
Strike =
r_T
v_T
b_T
T
F
a_F
c_F
F
R
d_F
b_F
17•6
c_T
T
R
a_T
b_T
T
F
c_F
Strike =
a_F
b_F
17•9
2 3 4 5 6
7 8 9 10 ½ Scale
Date
LINES IN A PLANE
Name
17•—

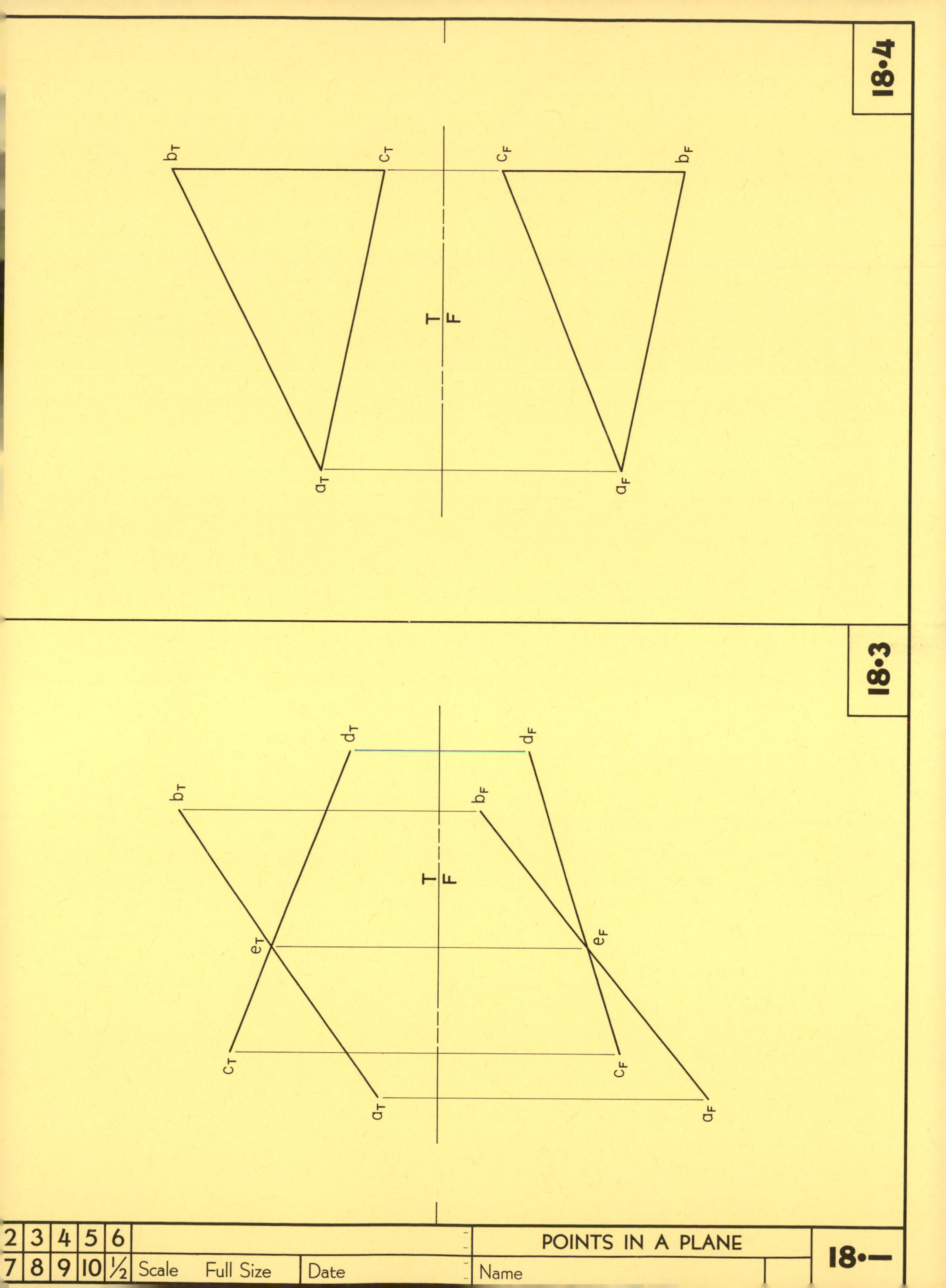
18·4
18·3
T
F
b_T
c_T
a_T
c_F
b_F
a_F
d_T
e_T
d_F
e_F
2 3 4 5 6
7 8 9 10 ½
Scale Full Size
Date
POINTS IN A PLANE
Name
18·—

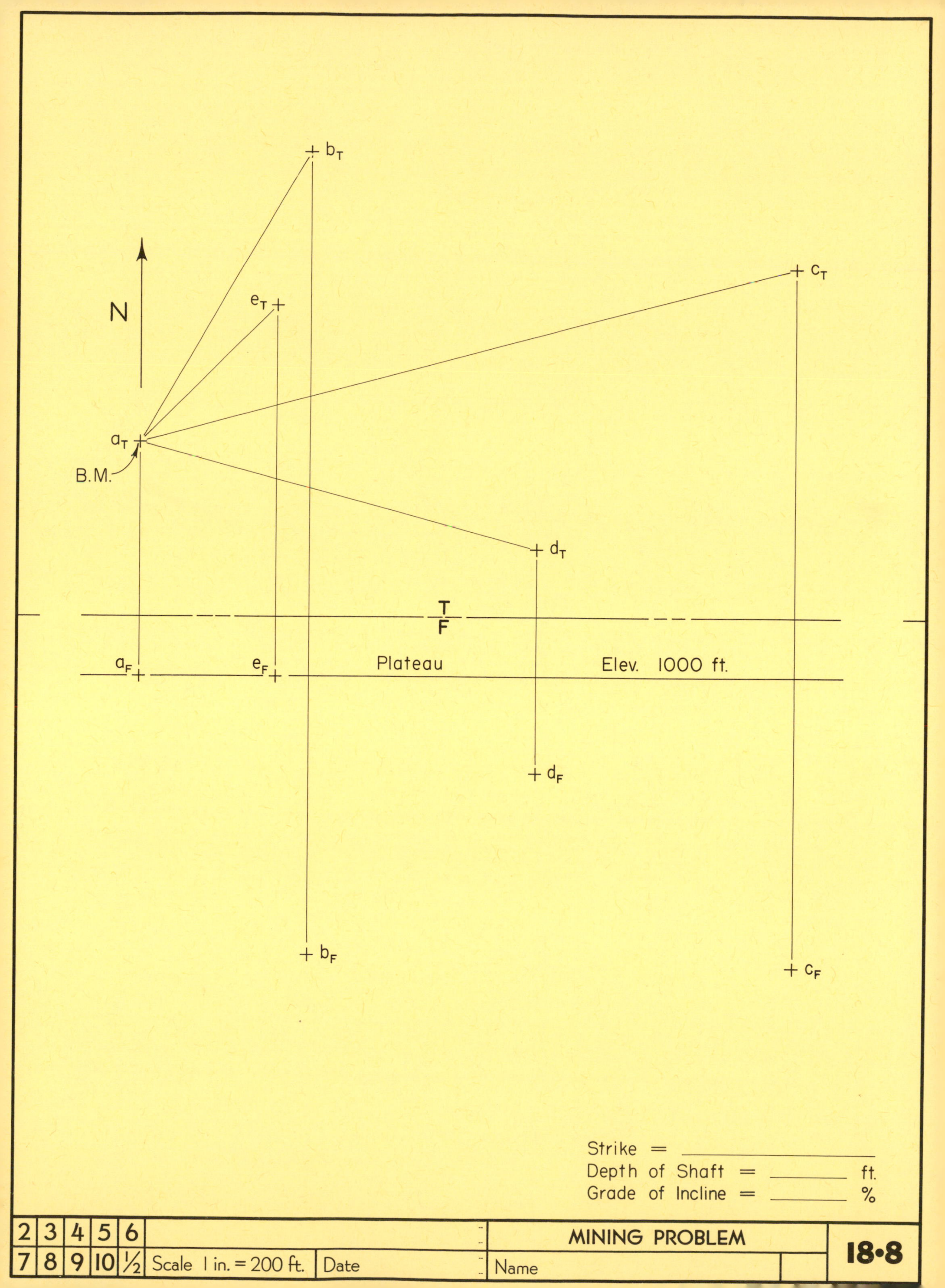

b_T
e_T
c_T
N
a_T
B.M.
d_T
T
F
a_F
e_F
Plateau
Elev. 1000 ft.
d_F
b_F
c_F
Strike =
Depth of Shaft = ft.
Grade of Incline = %
2 3 4 5 6
7 8 9 10 ½
Scale 1 in. = 200 ft.
Date
MINING PROBLEM
Name
18•8

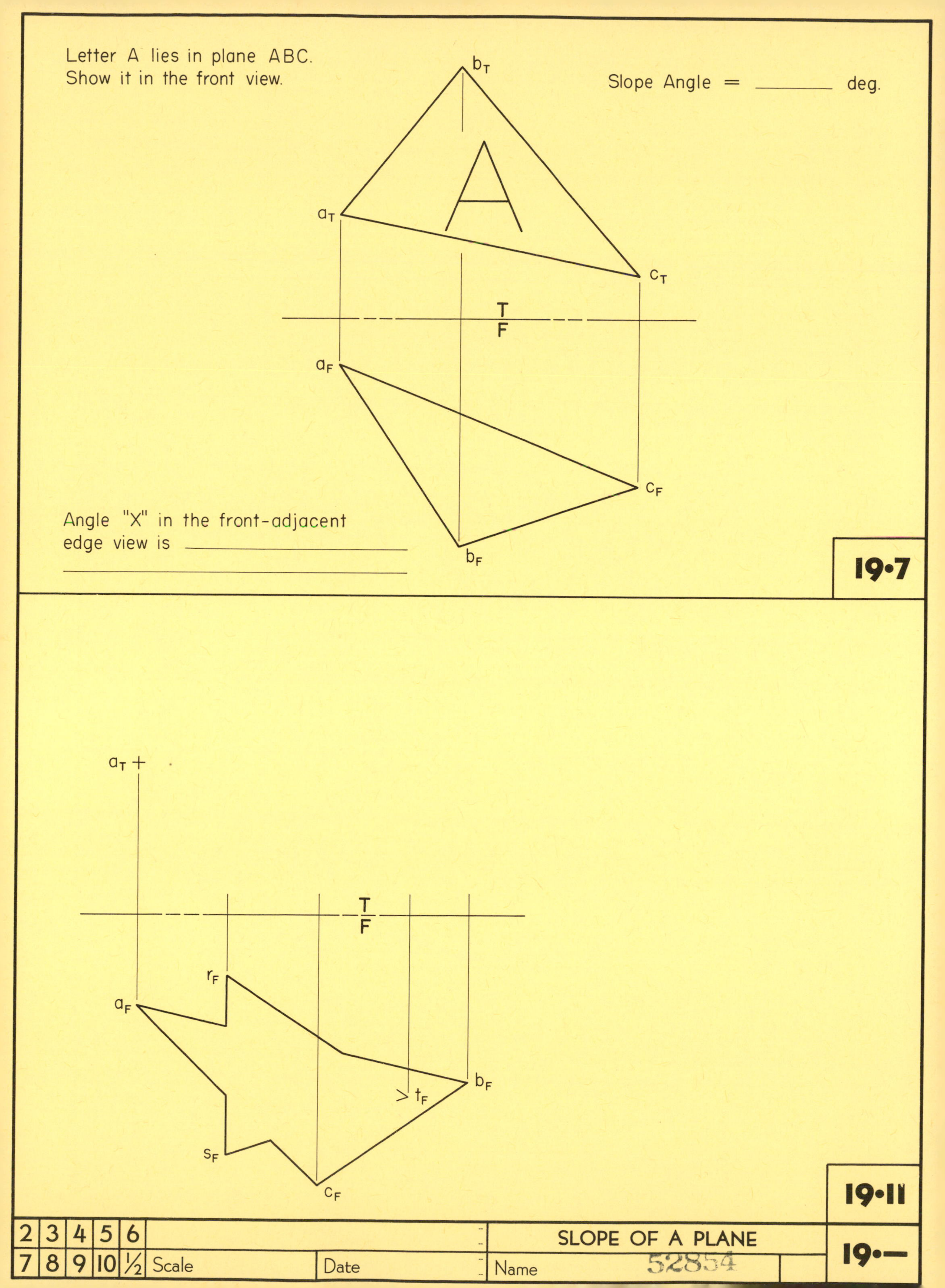
Letter A lies in plane ABC.
Show it in the front view.
Slope Angle = ________ deg.
b_T
a_T
c_T
T
F
a_F
c_F
b_F
Angle "X" in the front-adjacent
edge view is ________________
____
19•7
a_T
T
F
r_F
a_F
b_F
t_F
s_F
c_F
19•11
2 3 4 5 6
7 8 9 10 ½ Scale
Date
SLOPE OF A PLANE
Name
52854
19•—

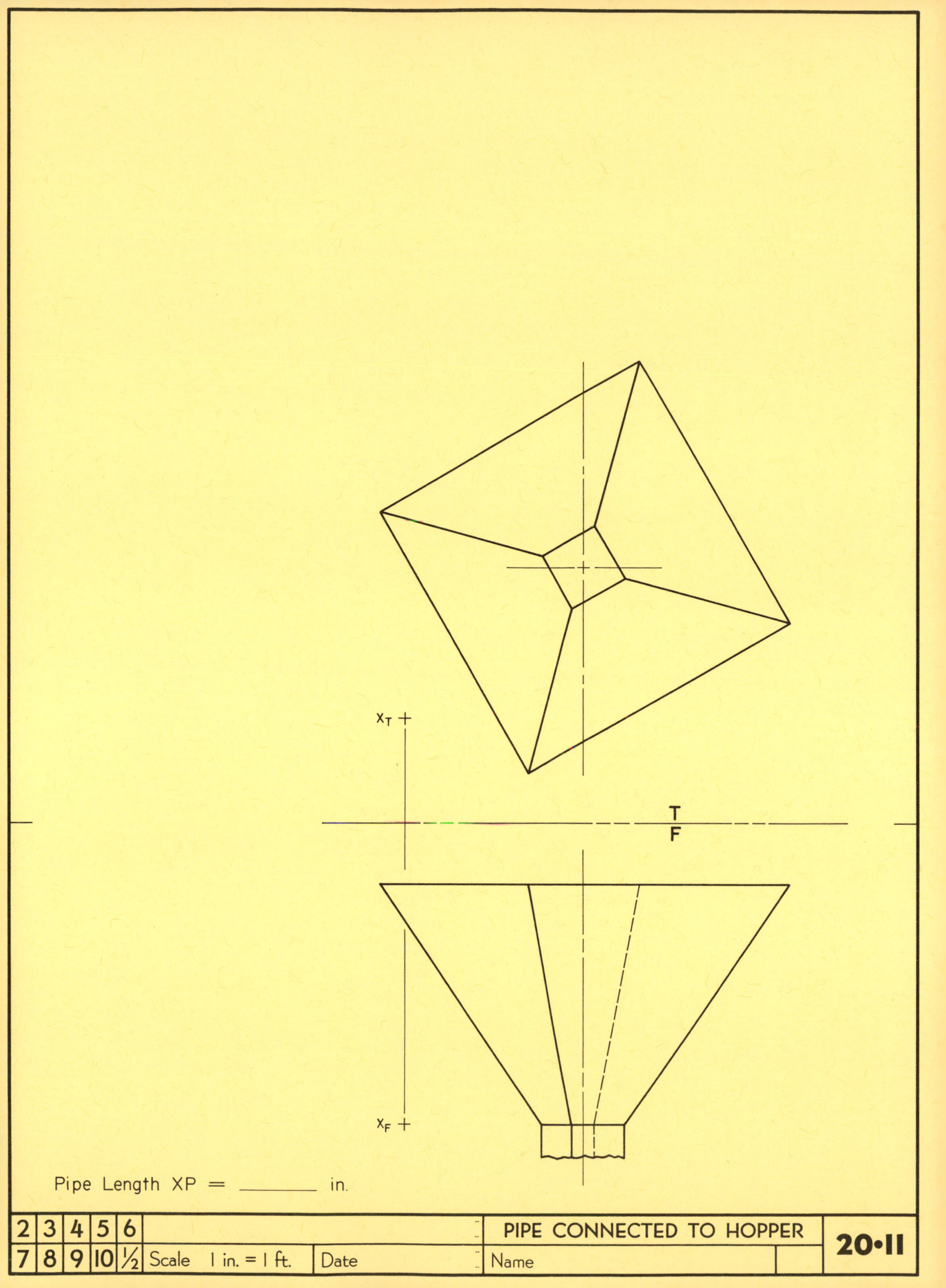

x_T
T
F
x_F
Pipe Length XP = ________ in.
2 3 4 5 6
7 8 9 10 ½
Scale 1 in. = 1 ft.
Date
PIPE CONNECTED TO HOPPER
Name
20•11

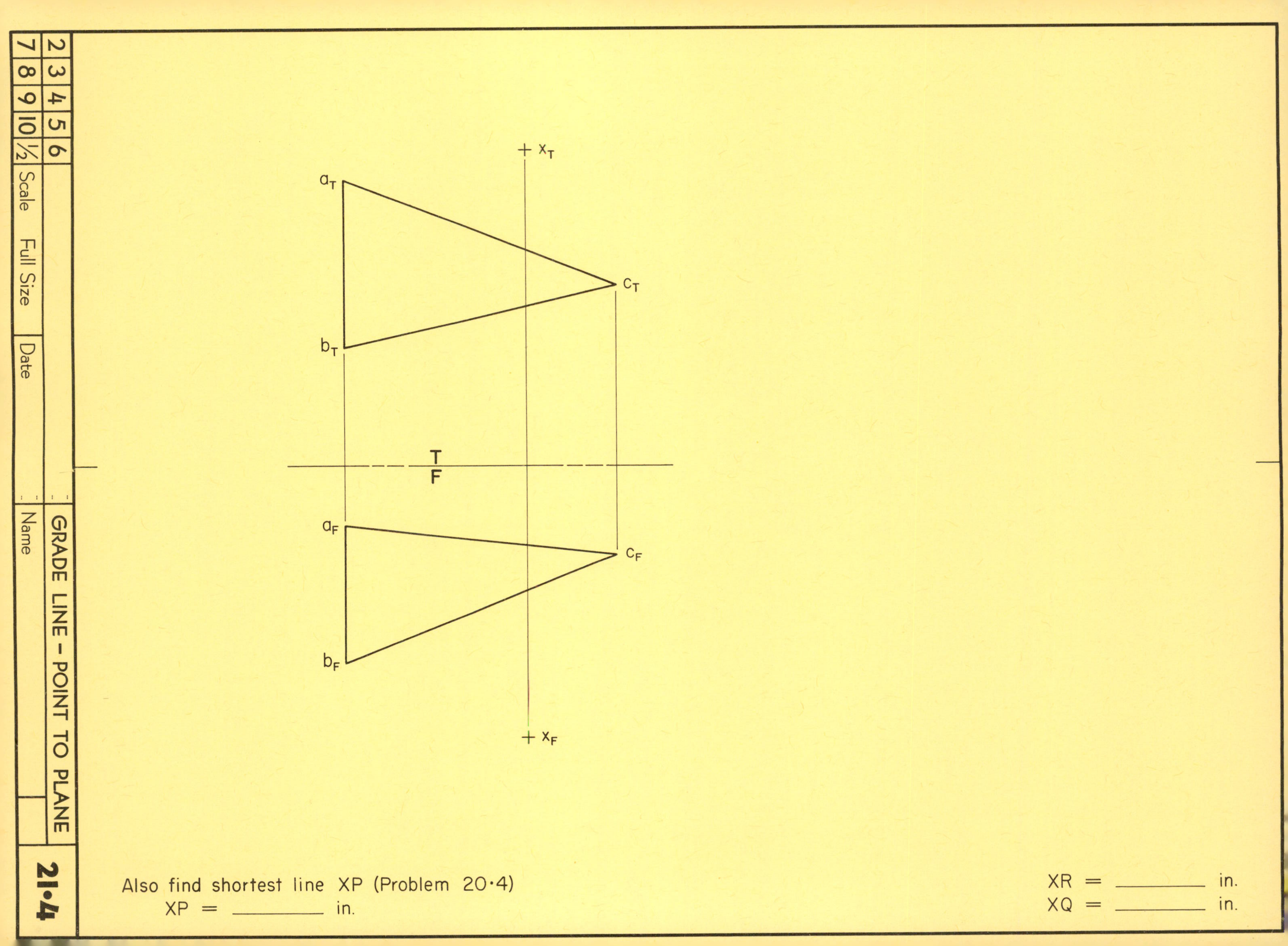
x_T
a_T
c_T
b_T
T
F
a_F
c_F
b_F
x_F
Also find shortest line XP (Problem 20·4)
XP = ______ in.
XR = ______ in.
XQ = ______ in.
2 3 4 5 6
7 8 9 10 ½
Scale Full Size
Date
GRADE LINE – POINT TO PLANE
Name
21·4

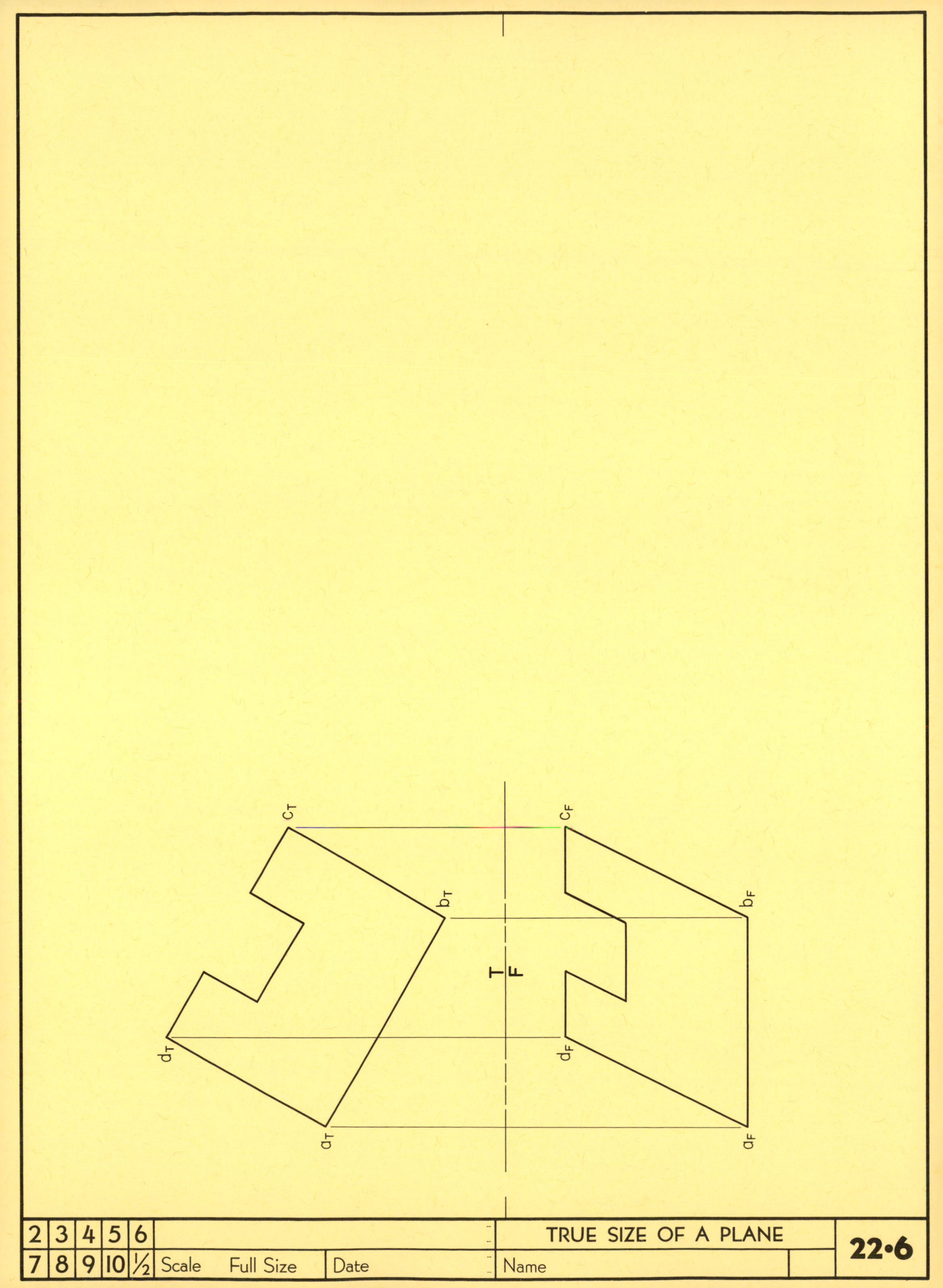

cT
bT
dT
aT
cF
bF
dF
aF
T
F
2 3 4 5 6
7 8 9 10 ½
Scale Full Size
Date
TRUE SIZE OF A PLANE
Name
22·6

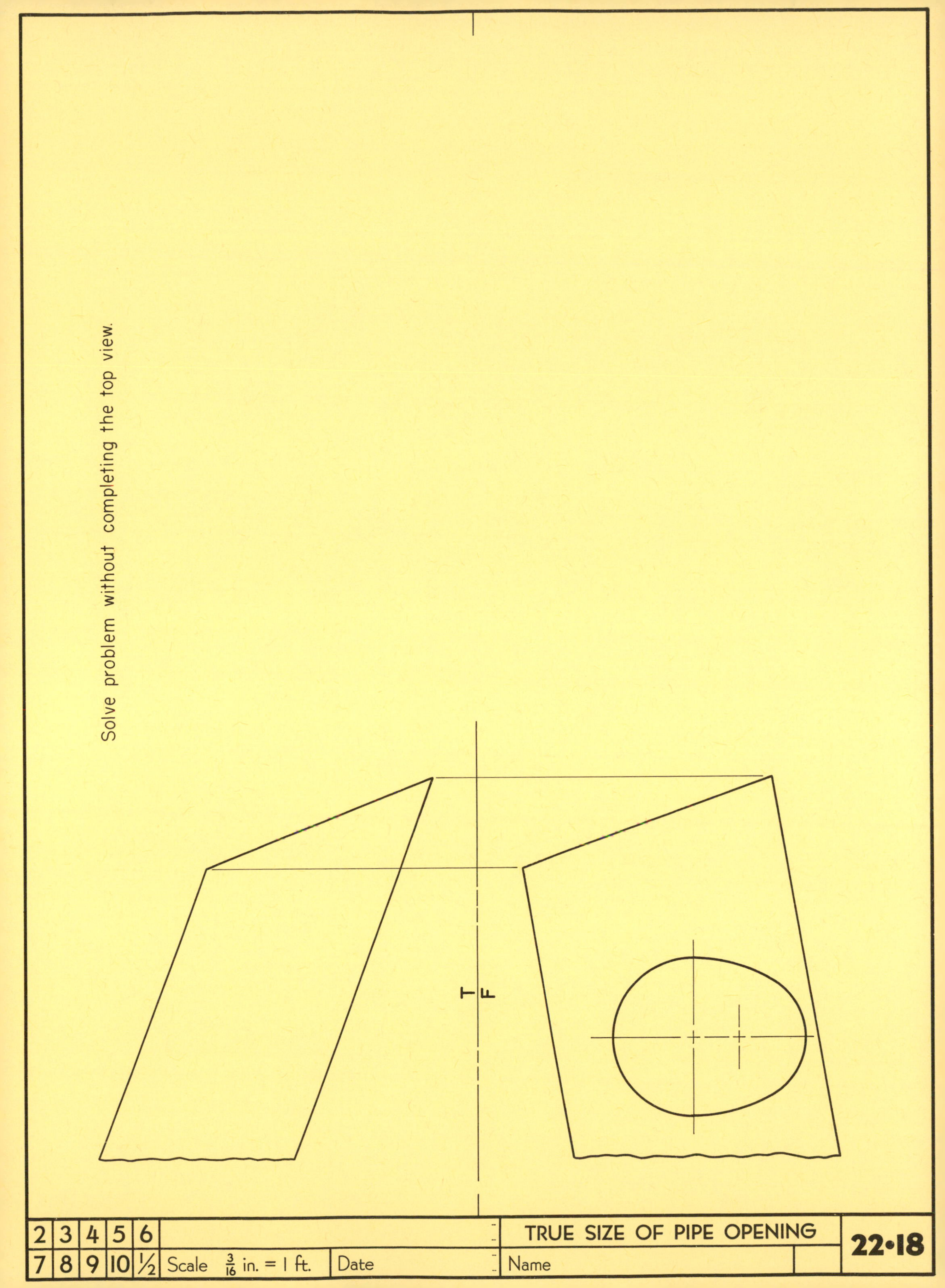
Solve problem without completing the top view.
T
F
2 3 4 5 6
7 8 9 10 ½
Scale 3/16 in. = 1 ft.
Date
TRUE SIZE OF PIPE OPENING
Name
22·18

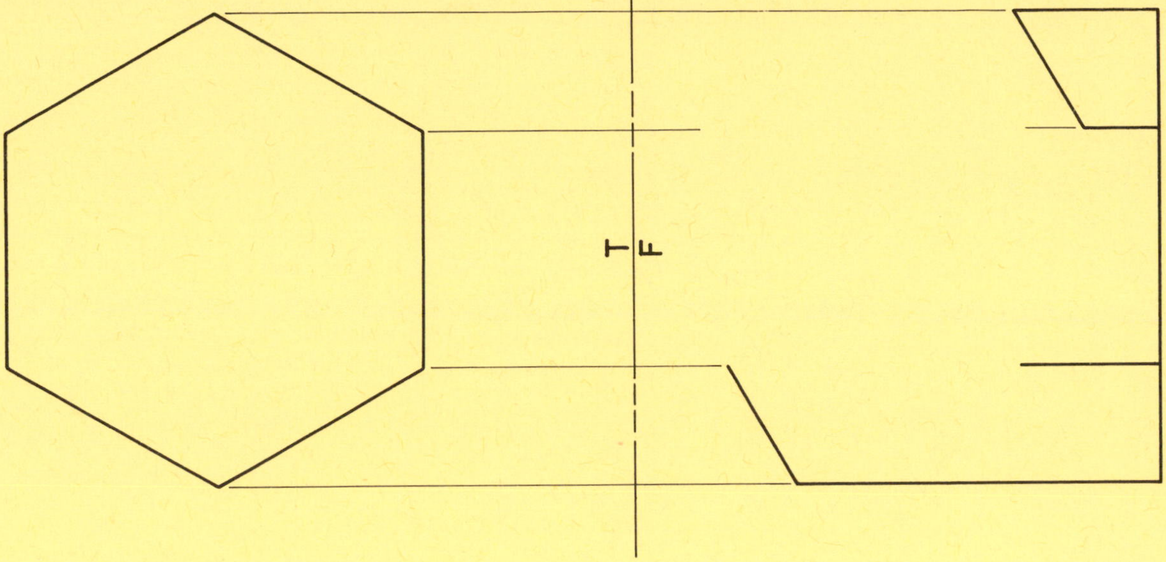

T
F

2	3	4	5	6		TRUE SIZE OF PRISM FACE	22•22
7	8	9	10	½	Scale / Date	Name	

Maximum Deflection Angles:

AB, position 1 = ______ deg.

AB, position 2 = ______ deg.

CD, position 1 = ______ deg.

CD, position 2 = ______ deg.

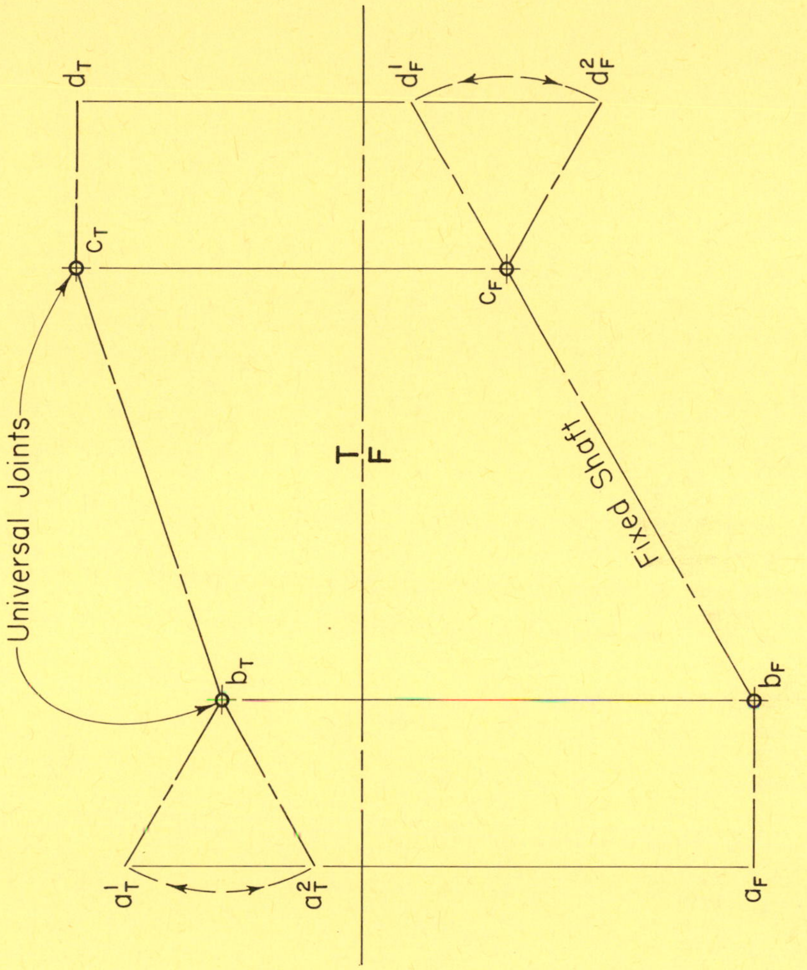

d_T

c_T

b_T

a_T^1 a_T^2

Universal Joints

T
F

d_F^1 d_F^2

c_F

Fixed Shaft

b_F

a_F

2	3	4	5	6		UNIVERSAL CONNECTED SHAFTS	23•10
7	8	9	10	½	Scale $1\frac{1}{2}$ in. = 1 ft. Date	Name	

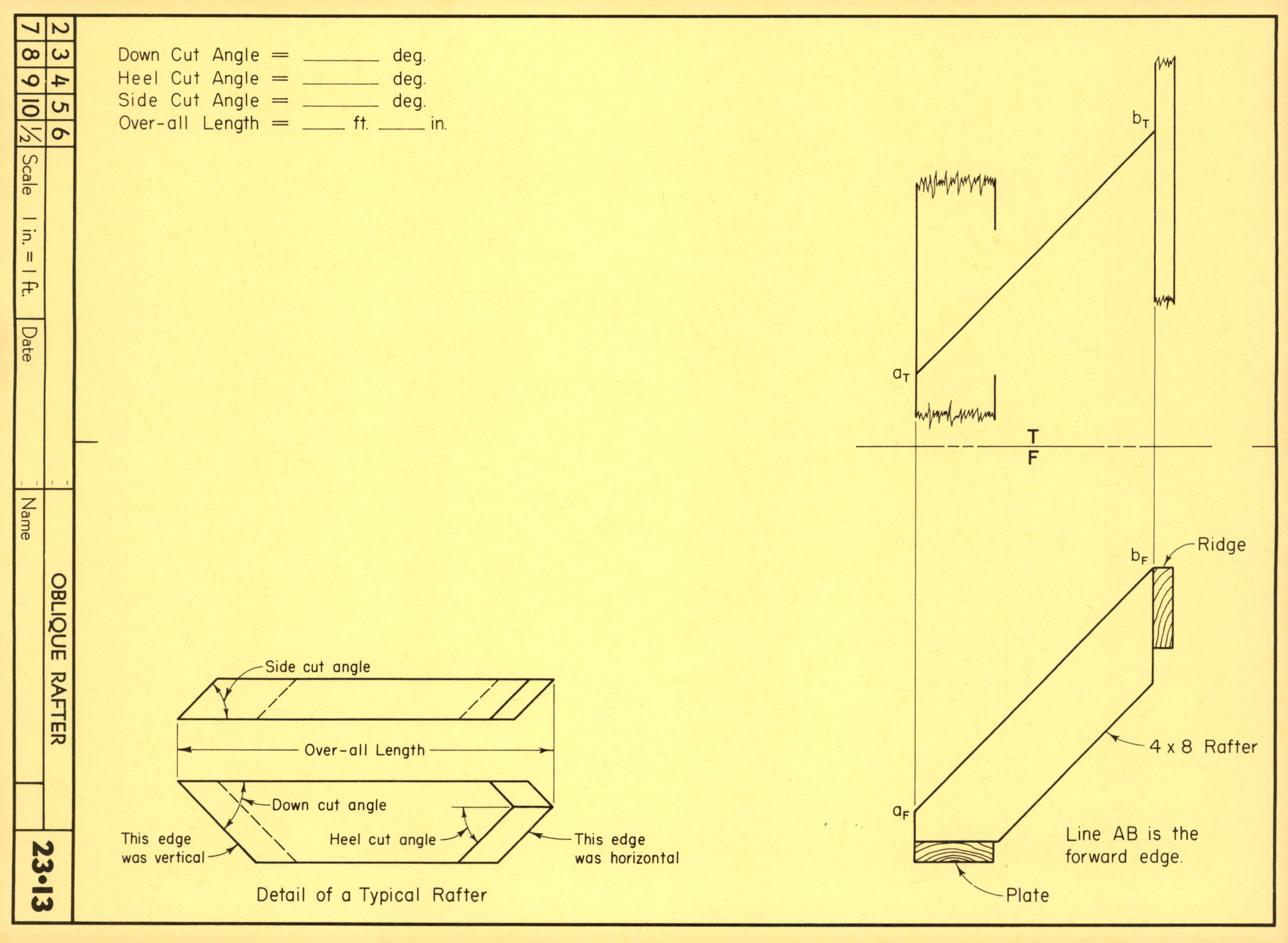

Down Cut Angle = ________ deg.
Heel Cut Angle = ________ deg.
Side Cut Angle = ________ deg.
Over-all Length = ____ ft. _____ in.
b_T
a_T
T
F
b_F
Ridge
a_F
4 x 8 Rafter
Plate
Line AB is the forward edge.
Side cut angle
Over-all Length
Down cut angle
Heel cut angle
This edge was vertical
This edge was horizontal
Detail of a Typical Rafter
2 3 4 5 6
7 8 9 10 ½
Scale 1 in. = 1 ft.
Date
Name
OBLIQUE RAFTER
23•13

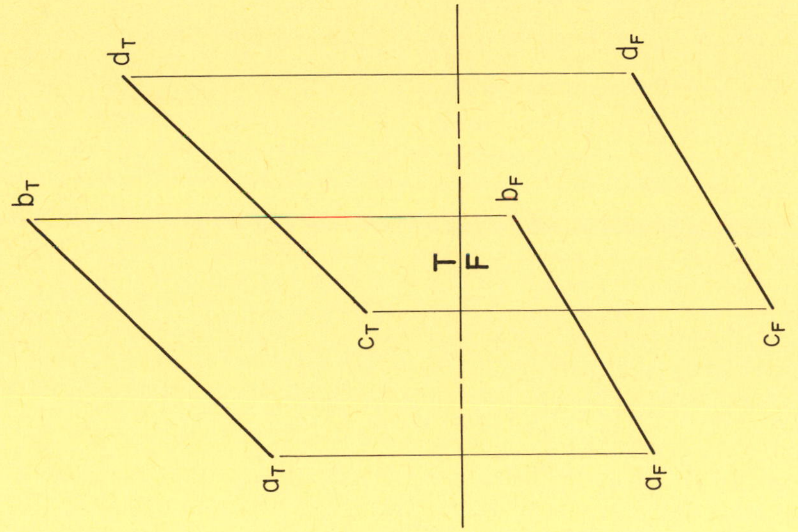

d_T d_F

b_T b_F

T

F

c_T c_F

a_T a_F

2	3	4	5	6		HEXAGON ON A PLANE	26·6
7	8	9	10	½	Scale Full Size / Date	Name	

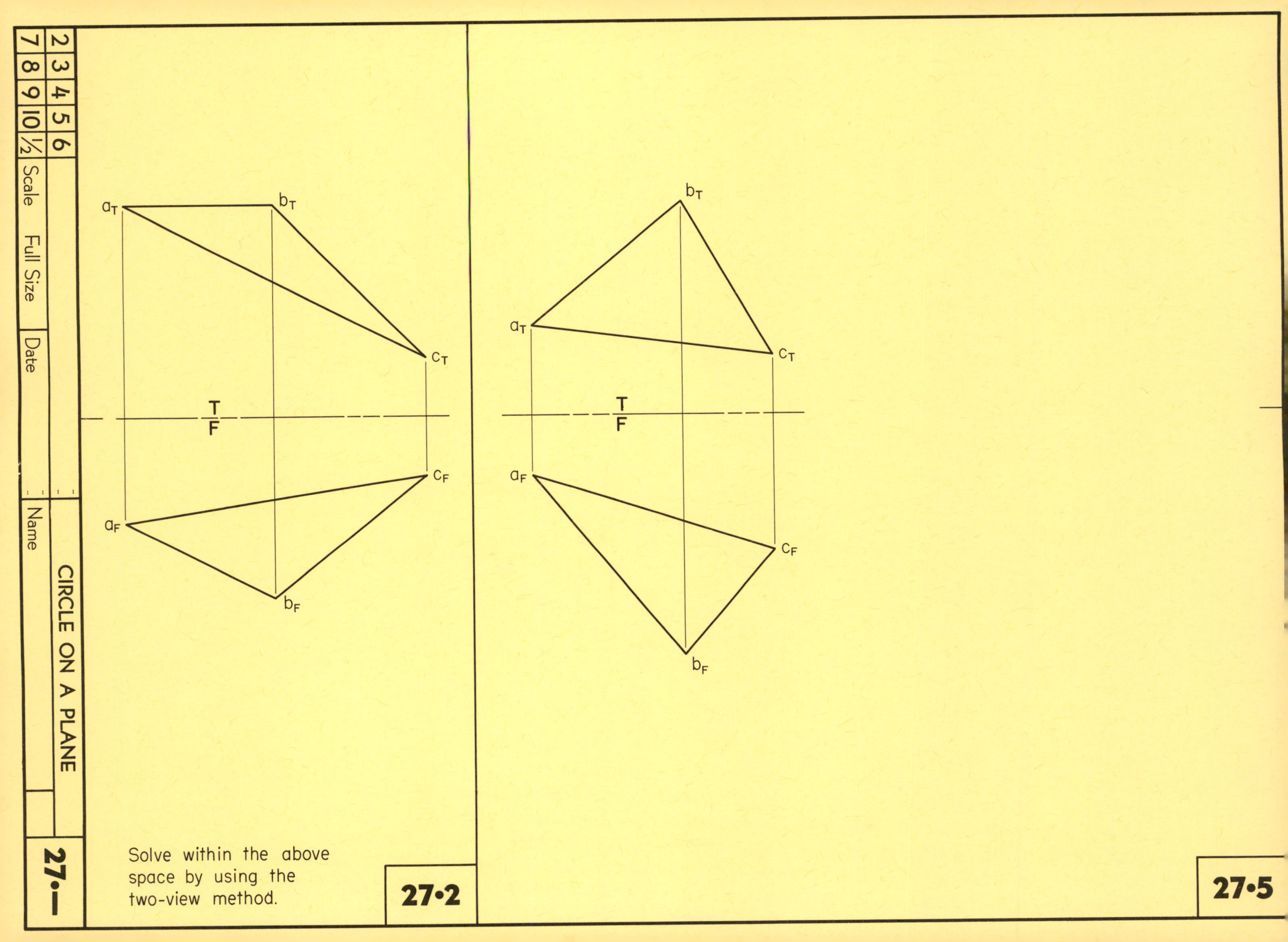

a_T
b_T
c_T
T
F
a_F
b_F
c_F
Solve within the above
space by using the
two-view method.
27•2
a_T
b_T
c_T
T
F
a_F
b_F
c_F
27•5
2 3 4 5 6
7 8 9 10 ½ Scale Full Size
Date
CIRCLE ON A PLANE
Name
27•—

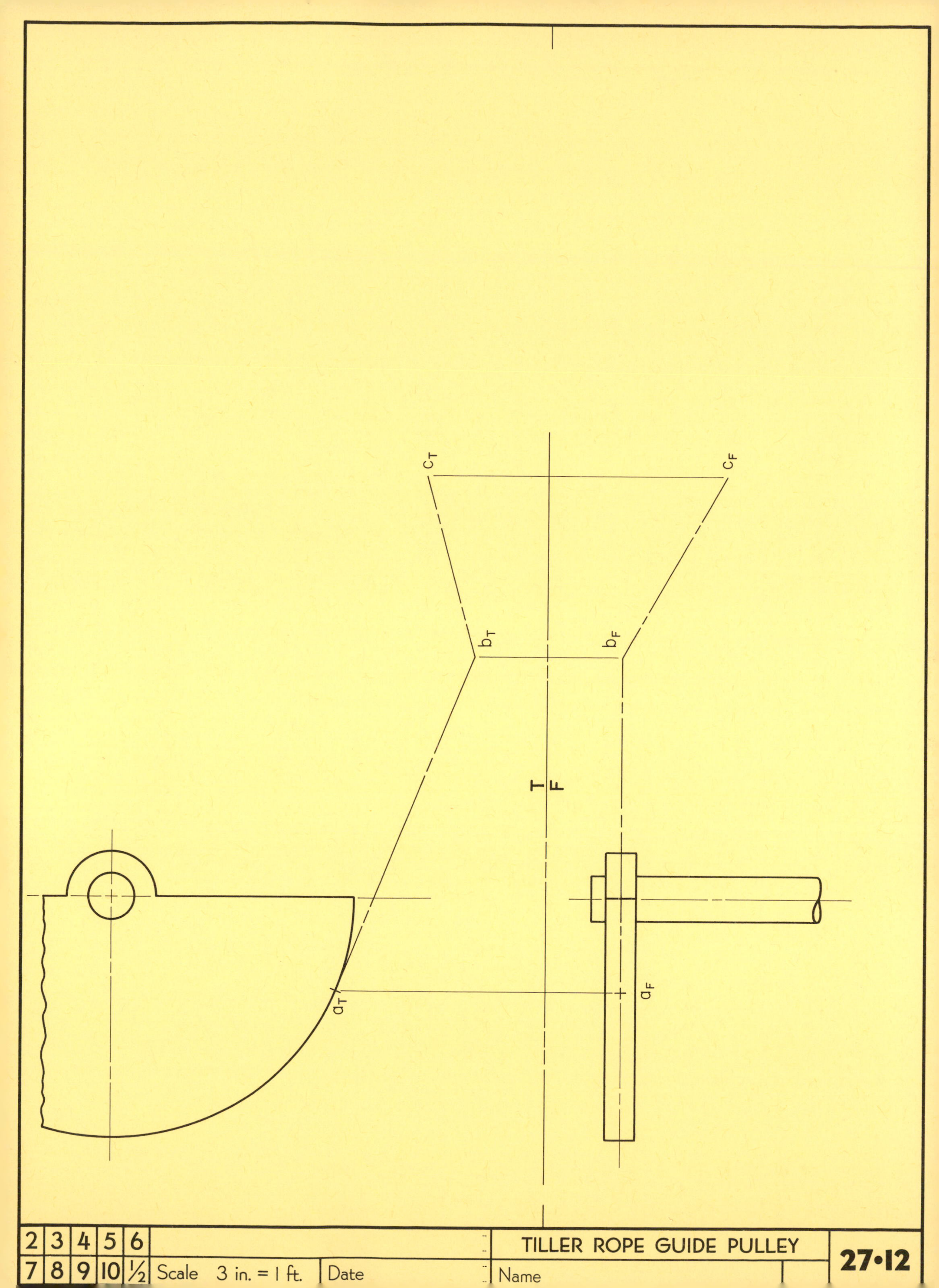

c_T
c_F
b_T
b_F
T
F
a_T
a_F
2 3 4 5 6
7 8 9 10 ½
Scale 3 in. = 1 ft.
Date
TILLER ROPE GUIDE PULLEY
Name
27•12

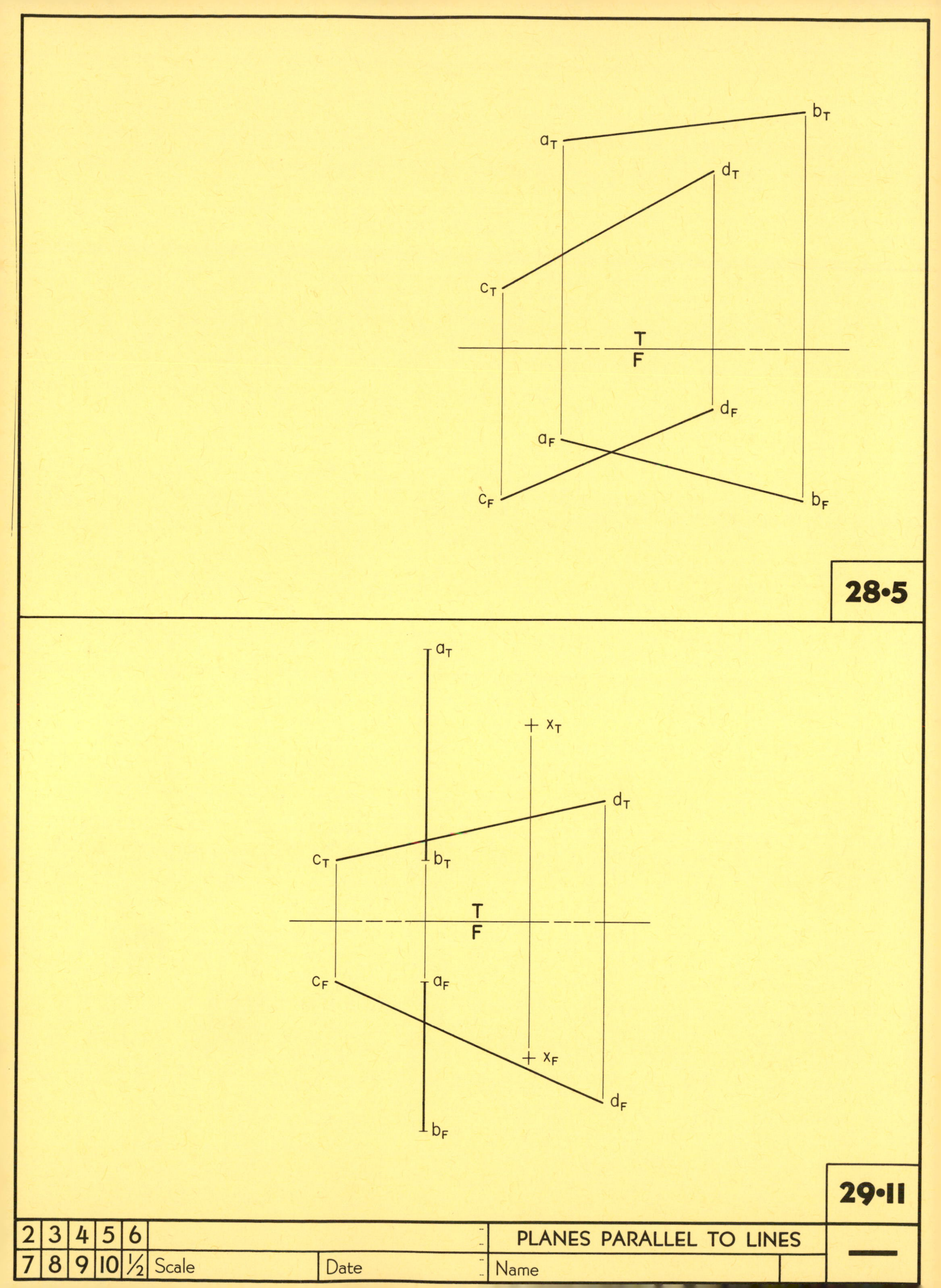

a_T
b_T
d_T
c_T
T
F
d_F
a_F
c_F
b_F
28•5
a_T
x_T
d_T
c_T
b_T
T
F
c_F
a_F
x_F
d_F
b_F
29•11
2 3 4 5 6
7 8 9 10 ½
Scale
Date
Name
PLANES PARALLEL TO LINES

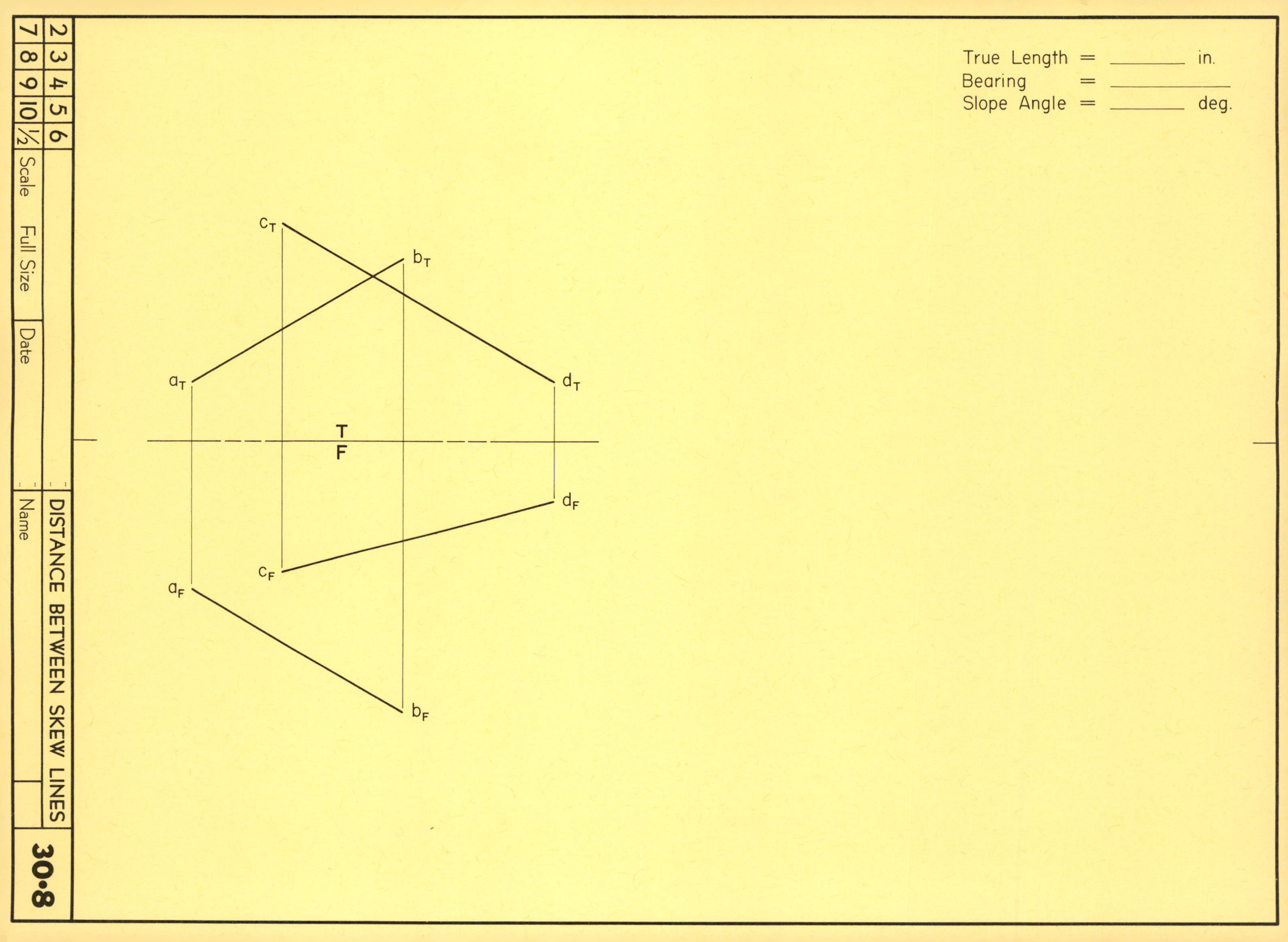

True Length = ______ in.
Bearing = ______
Slope Angle = ______ deg.
c_T
b_T
a_T
d_T
T
F
d_F
c_F
a_F
b_F
2 3 4 5 6
7 8 9 10 ½
Scale Full Size
Date
Name
DISTANCE BETWEEN SKEW LINES
30•8

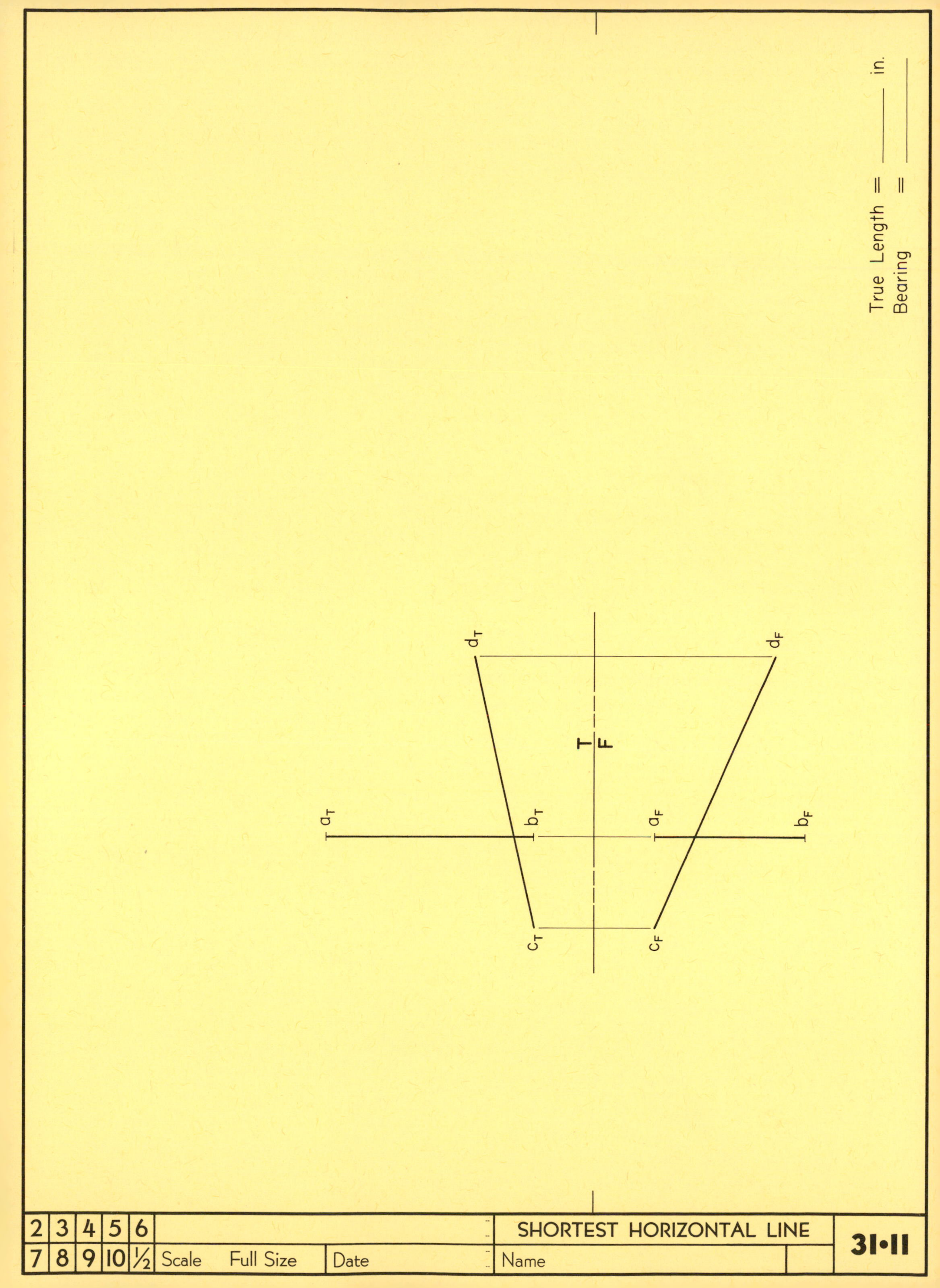

True Length = ______ in.
Bearing = ______
d_T
d_F
T
F
a_T
b_T
a_F
b_F
c_T
c_F
2 3 4 5 6
7 8 9 10 ½
Scale Full Size
Date
SHORTEST HORIZONTAL LINE
Name
31•11

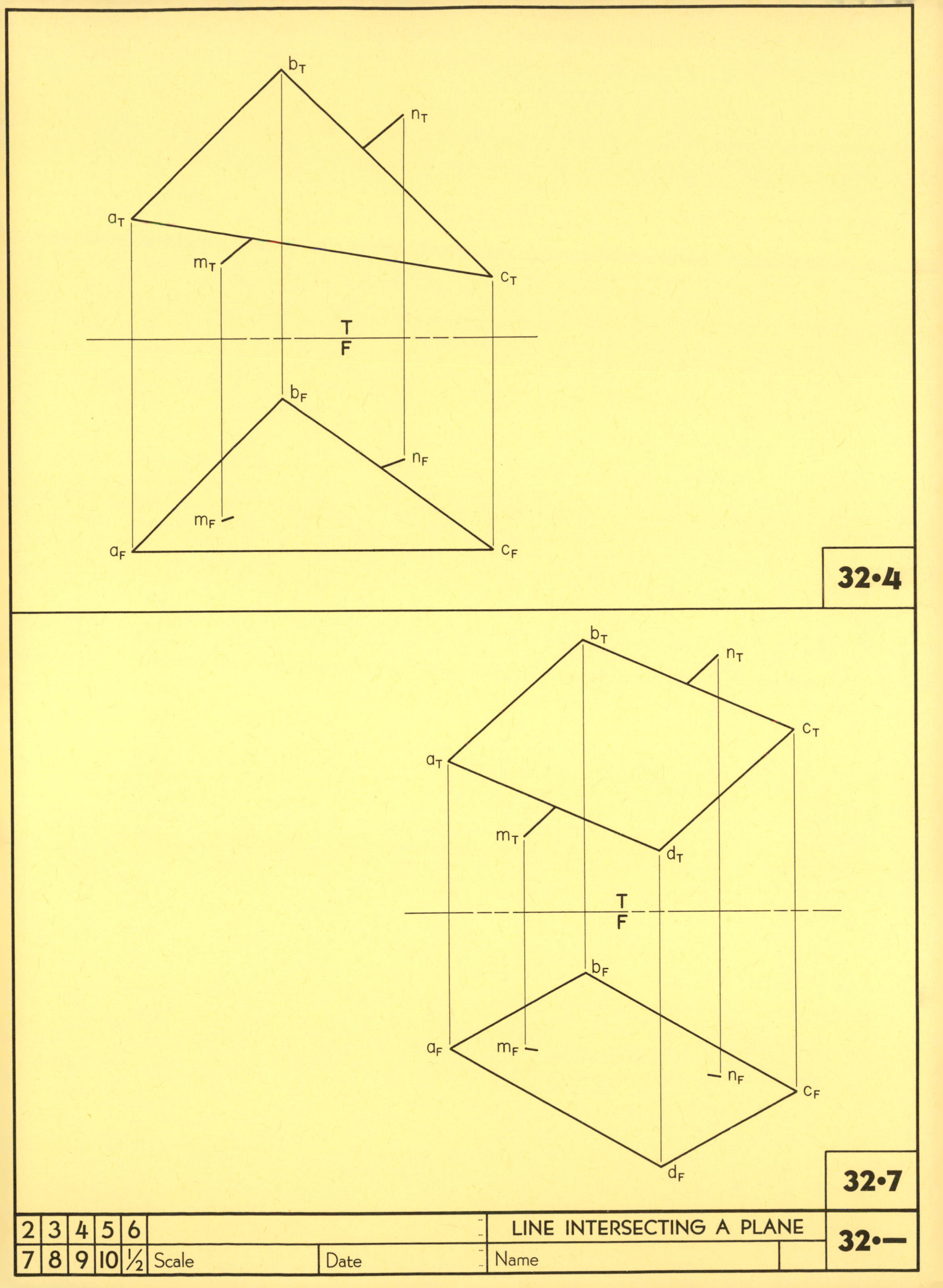

b_T
n_T
a_T
m_T
c_T
T
F
b_F
n_F
m_F
a_F
c_F
32•4
b_T
n_T
c_T
a_T
m_T
d_T
T
F
b_F
a_F
m_F
n_F
c_F
d_F
32•7
2 3 4 5 6
7 8 9 10 ½
Scale
Date
LINE INTERSECTING A PLANE
Name
32•—

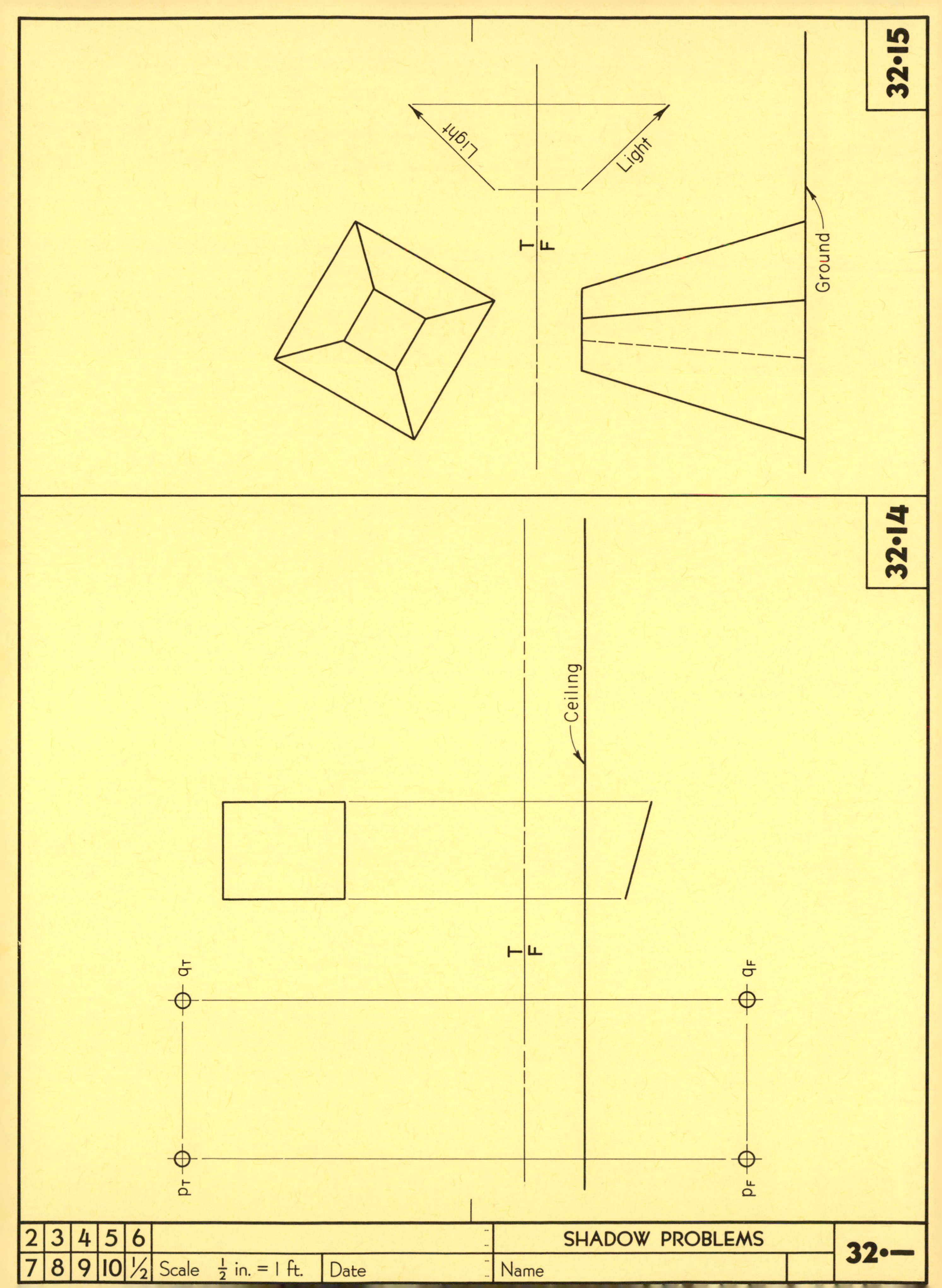
32·15
Light
Light
T
F
Ground
32·14
Ceiling
T
F
q_T
q_F
p_T
p_F
2 3 4 5 6
7 8 9 10 ½
Scale ½ in. = 1 ft.
Date
SHADOW PROBLEMS
Name
32•—

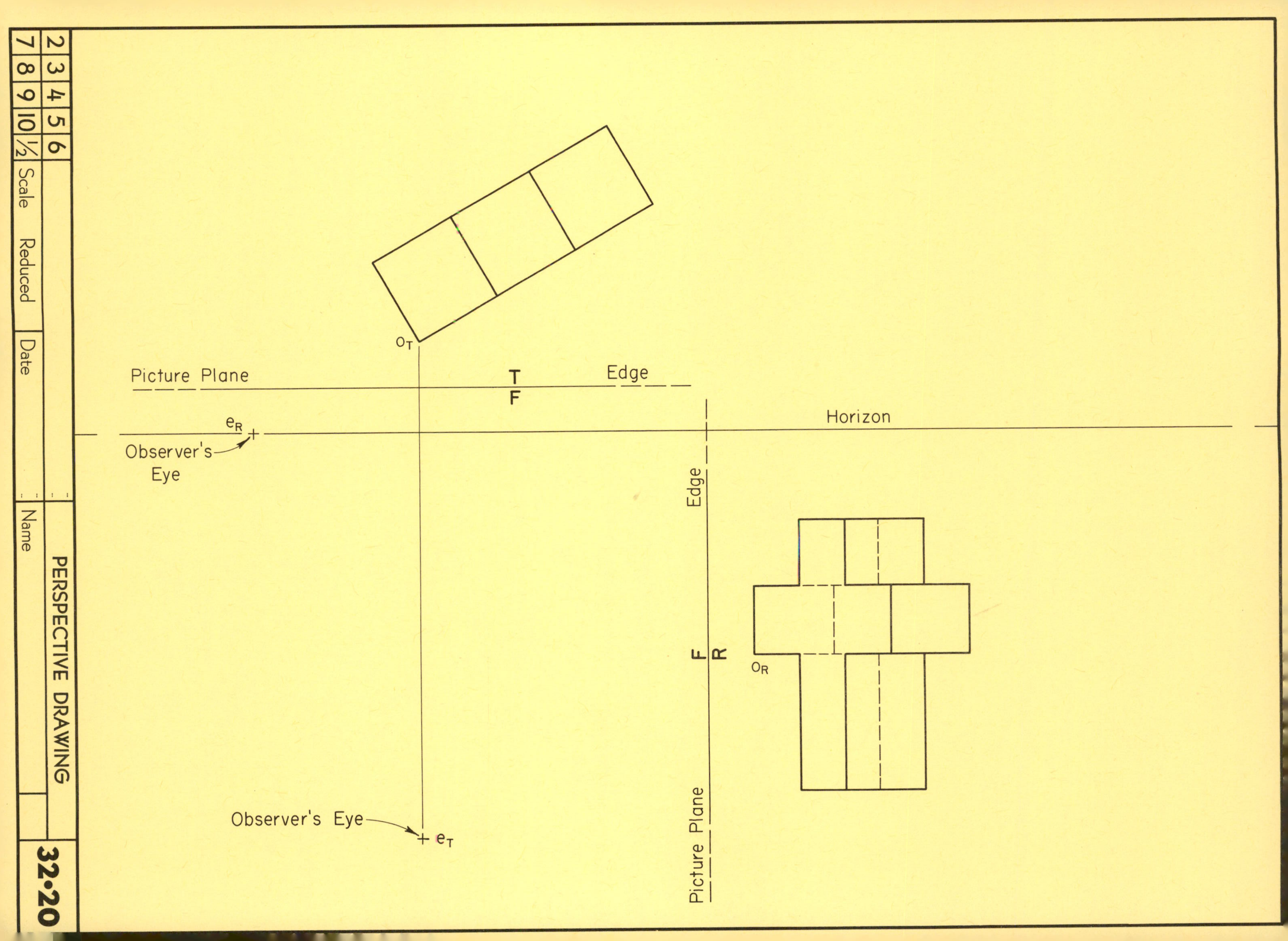

Picture Plane
T
F
Edge
O_T
e_R
Observer's Eye
Horizon
Edge
F
R
O_R
Picture Plane
Observer's Eye
e_T
2 3 4 5 6
7 8 9 10 ½
Scale Reduced
Date
Name
PERSPECTIVE DRAWING
32•20

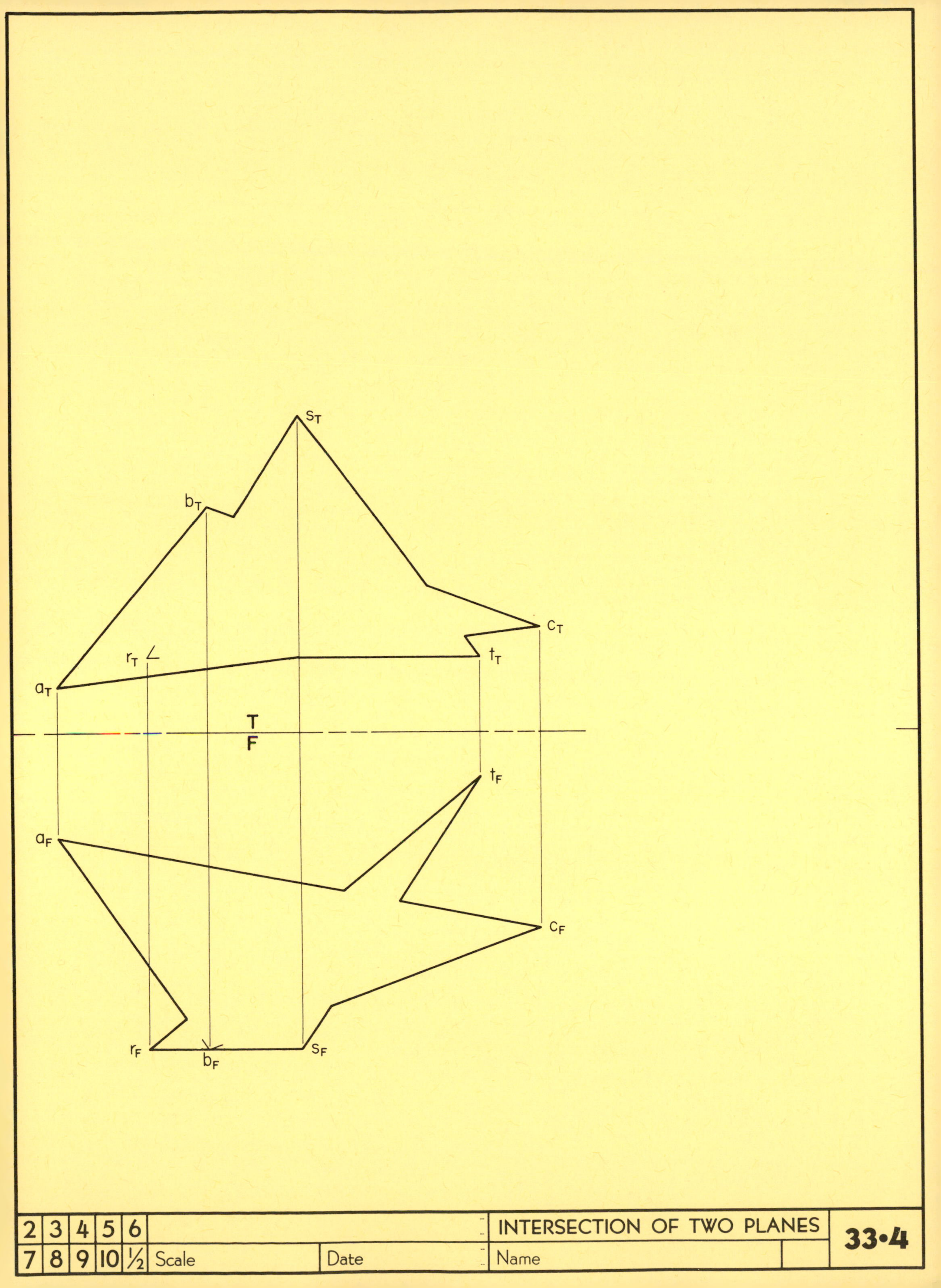

s_T
b_T
c_T
r_T
t_T
a_T
T
F
t_F
a_F
c_F
r_F
b_F
s_F
2 3 4 5 6
7 8 9 10 ½ Scale
Date
INTERSECTION OF TWO PLANES
Name
33•4

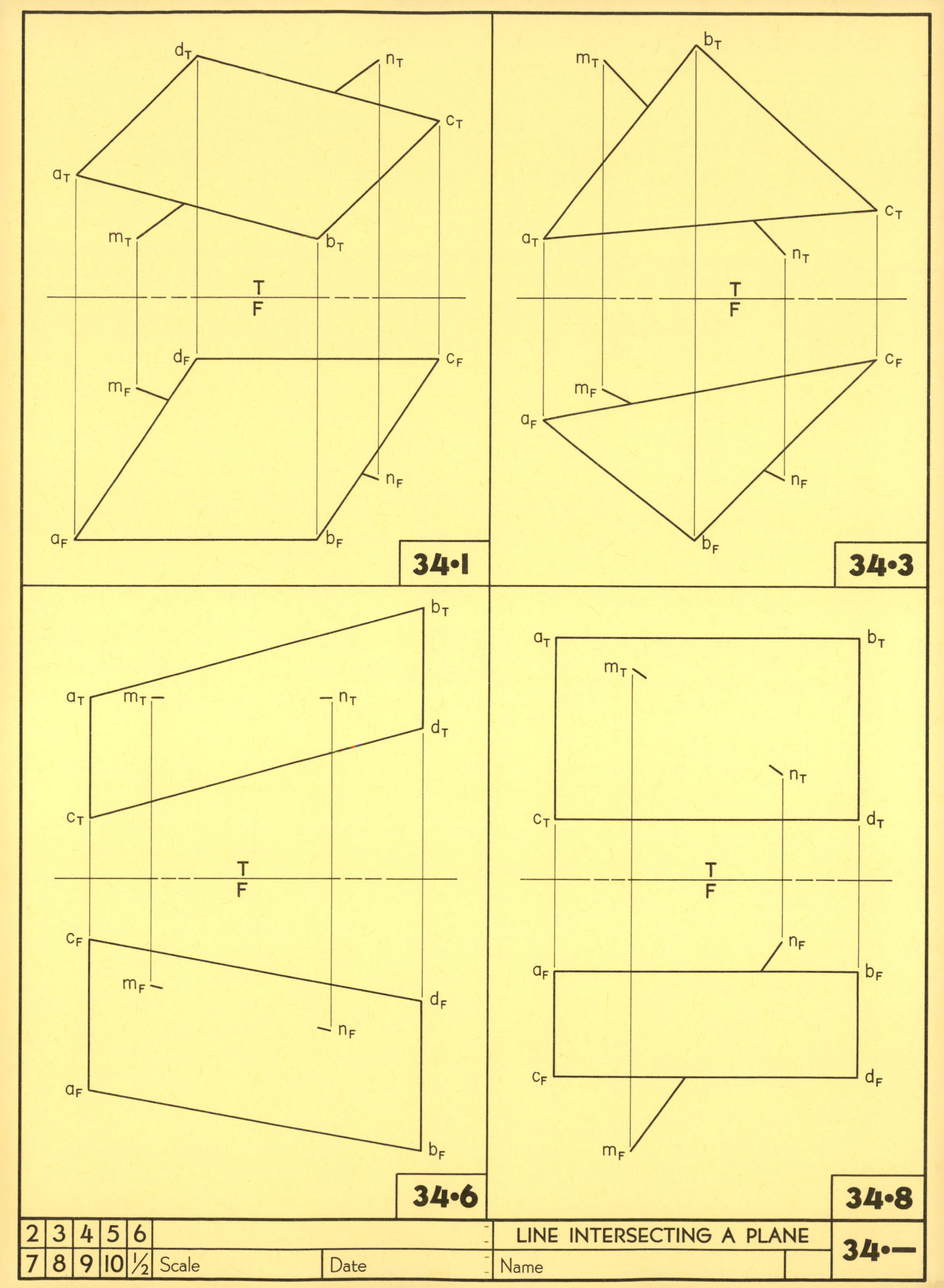

34•1
34•3
34•6
34•8
2 3 4 5 6
7 8 9 10 ½
Scale
Date
LINE INTERSECTING A PLANE
Name
34•—

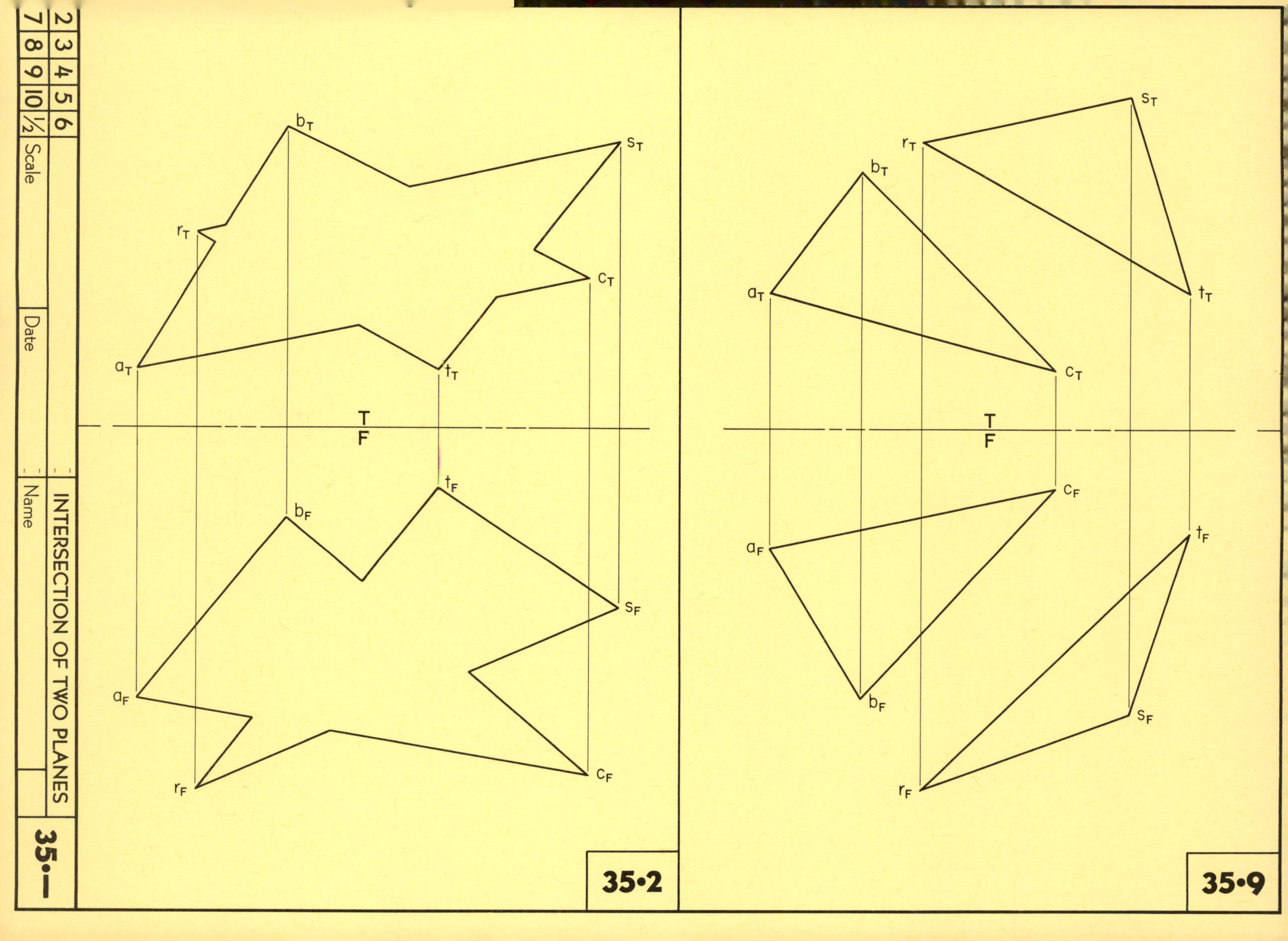
b_T
s_T
r_T
c_T
a_T
t_T
T
F
t_F
b_F
s_F
a_F
r_F
c_F
35•2
s_T
r_T
b_T
a_T
t_T
c_T
T
F
c_F
t_F
a_F
b_F
s_F
r_F
35•9
2 3 4 5 6
7 8 9 10 ½
Scale
Date
Name
INTERSECTION OF TWO PLANES
35•—

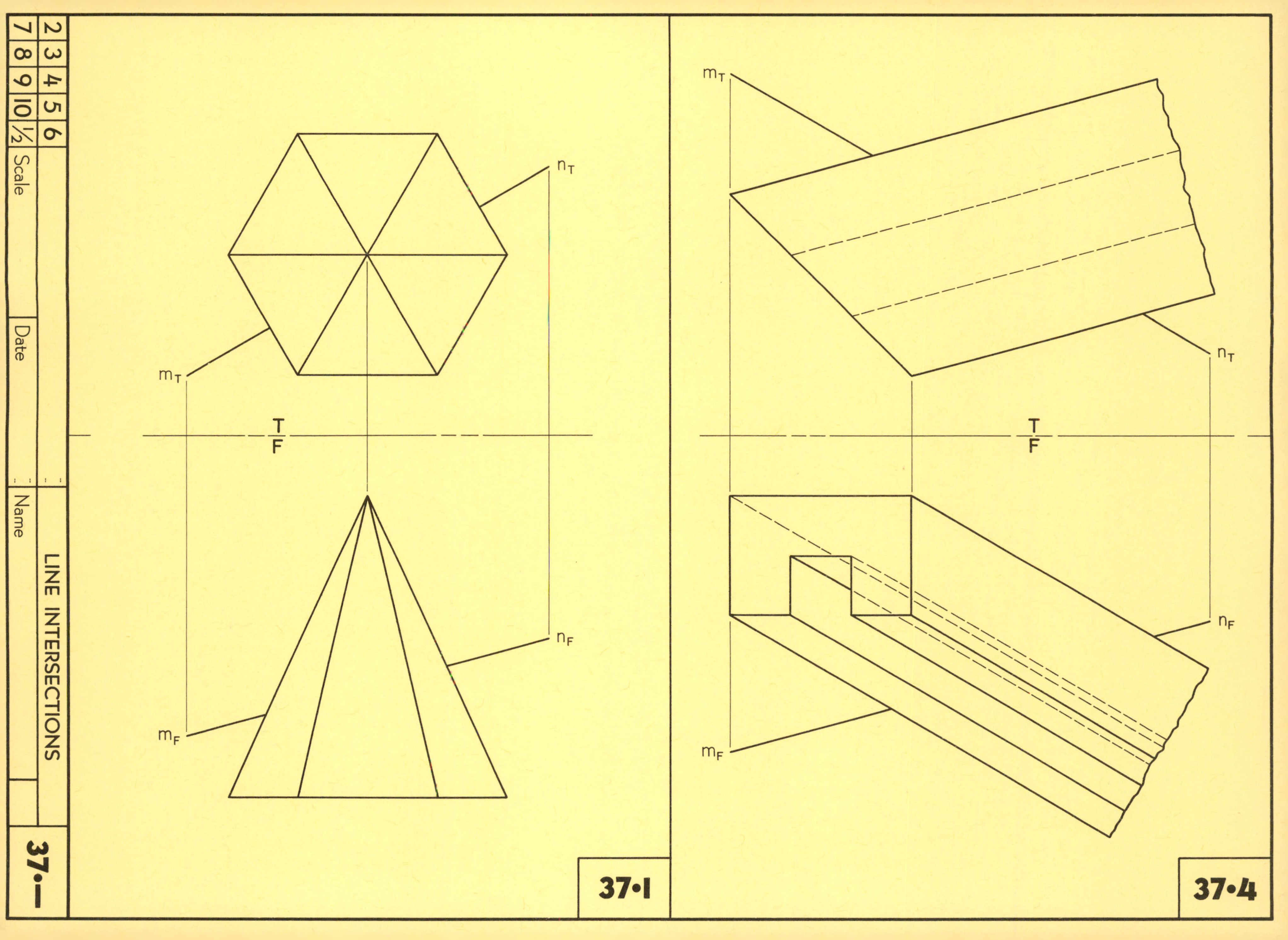

n_T
m_T
T
F
n_F
m_F
37•1
m_T
n_T
T
F
n_F
m_F
37•4
2 3 4 5 6
7 8 9 10 ½
Scale
Date
Name
LINE INTERSECTIONS
37•—

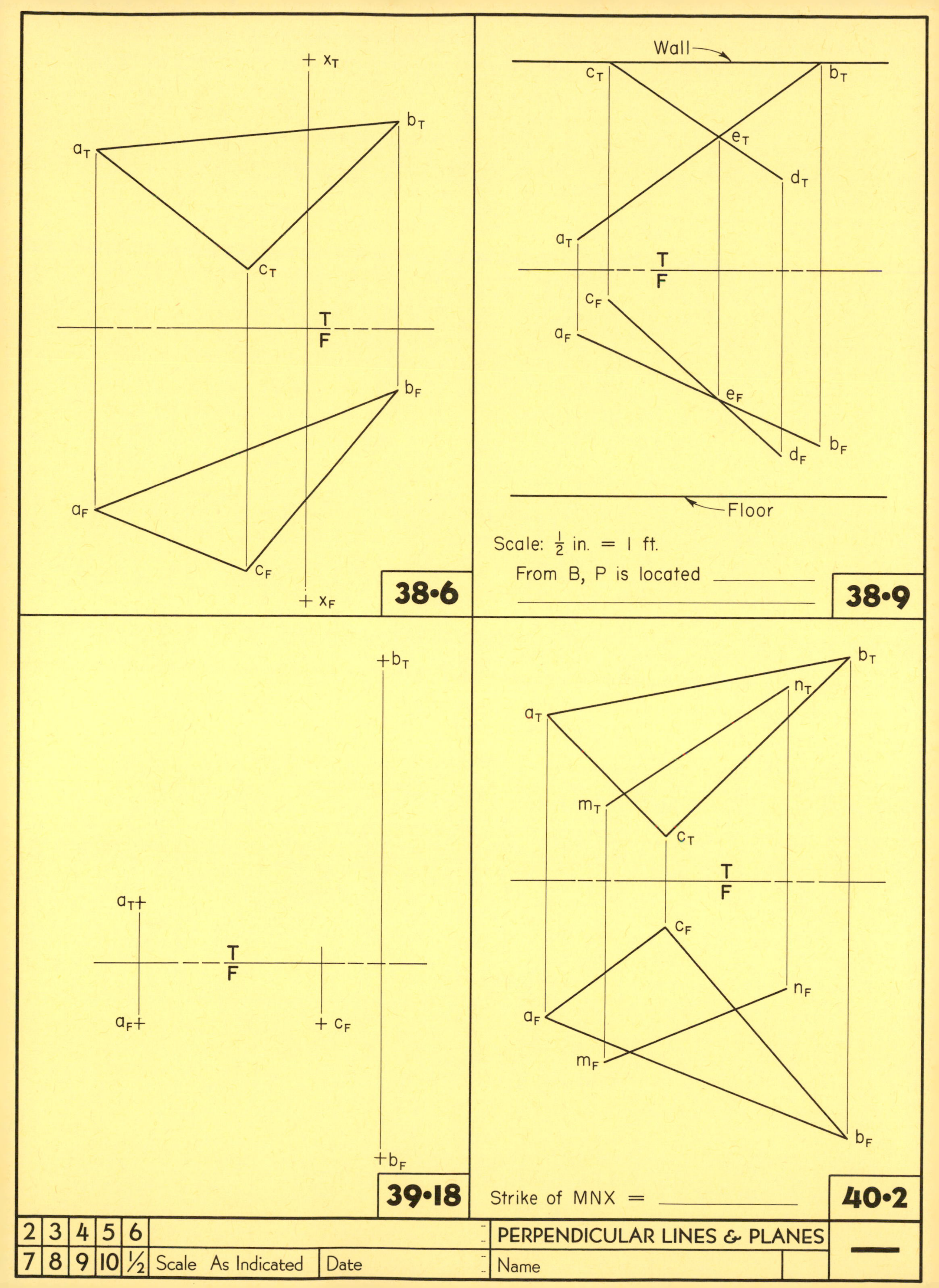
xT
aT
bT
cT
T
F
bF
aF
cF
xF
38•6
Wall
cT
bT
eT
dT
aT
T
F
cF
aF
eF
bF
dF
Floor
Scale: 1/2 in. = 1 ft.
From B, P is located
38•9
bT
aT
T
F
aF
cF
bF
39•18
bT
nT
aT
mT
cT
T
F
cF
nF
aF
mF
bF
Strike of MNX =
40•2
2 3 4 5 6
7 8 9 10 1/2
Scale As Indicated
Date
PERPENDICULAR LINES & PLANES
Name

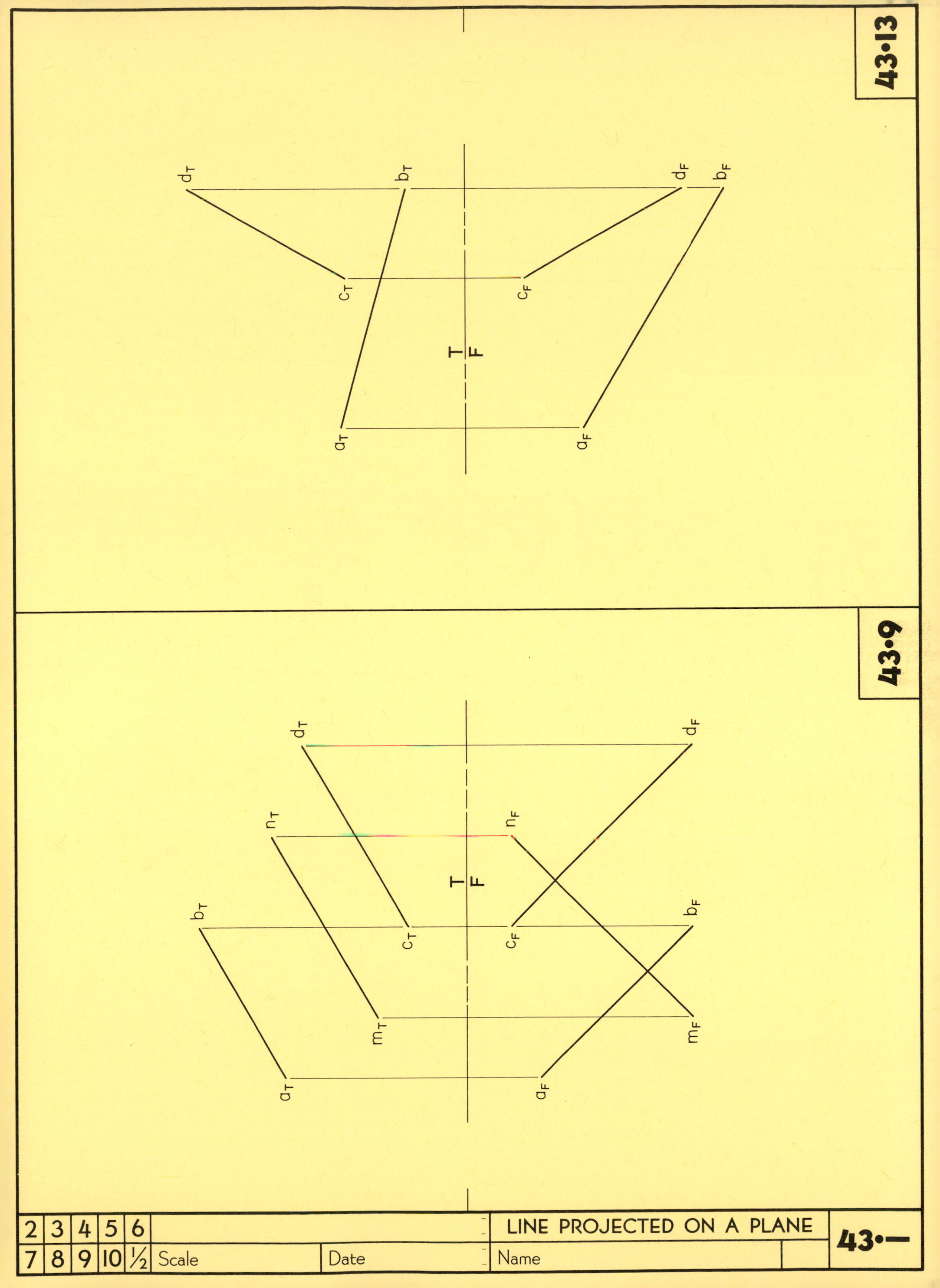

43·13
43·9
T
F
LINE PROJECTED ON A PLANE
43·—
2 3 4 5 6
7 8 9 10 ½
Scale
Date
Name

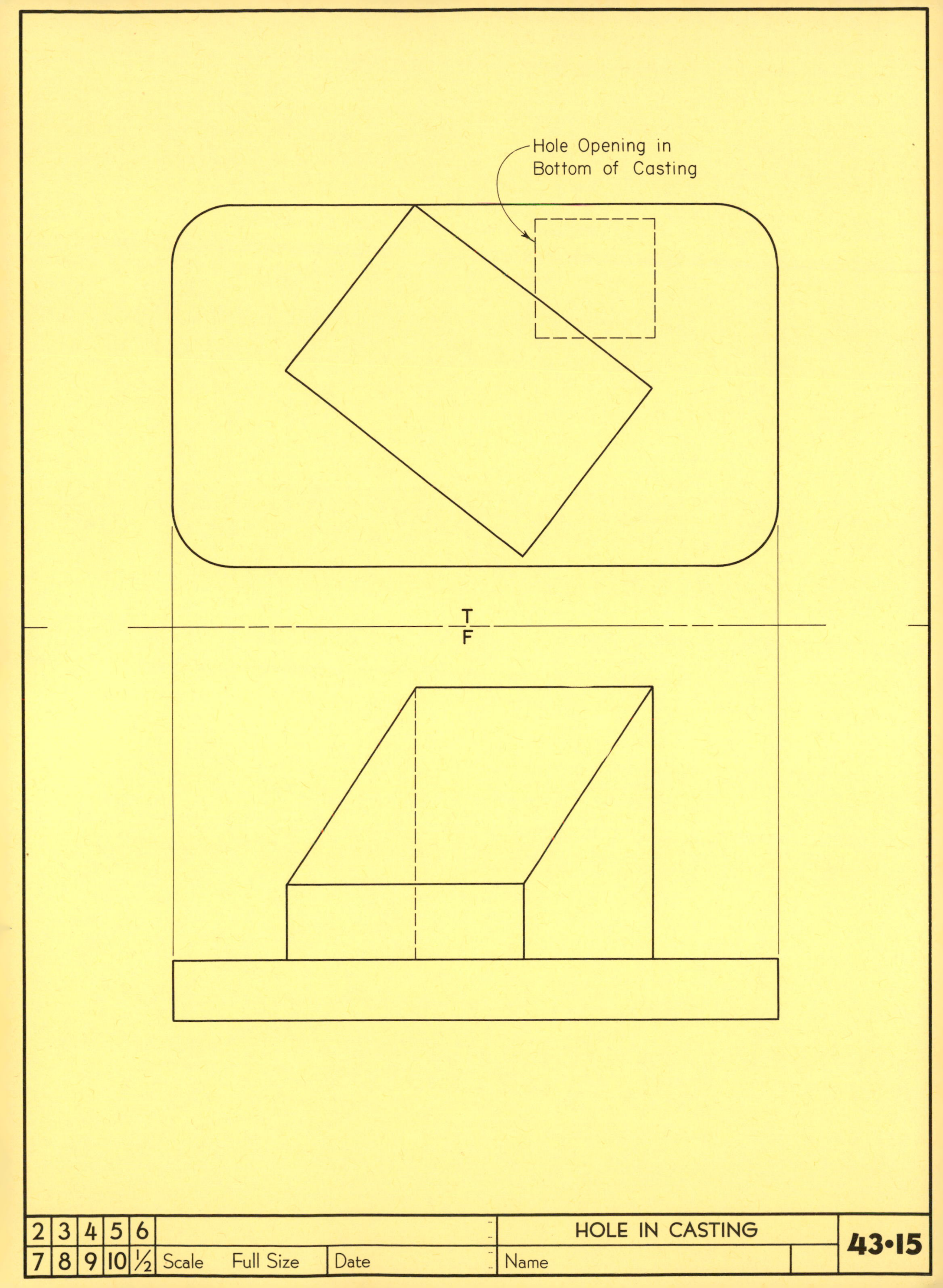
Hole Opening in
Bottom of Casting
T
F
2 3 4 5 6
7 8 9 10 ½
Scale Full Size
Date
HOLE IN CASTING
Name
43•15

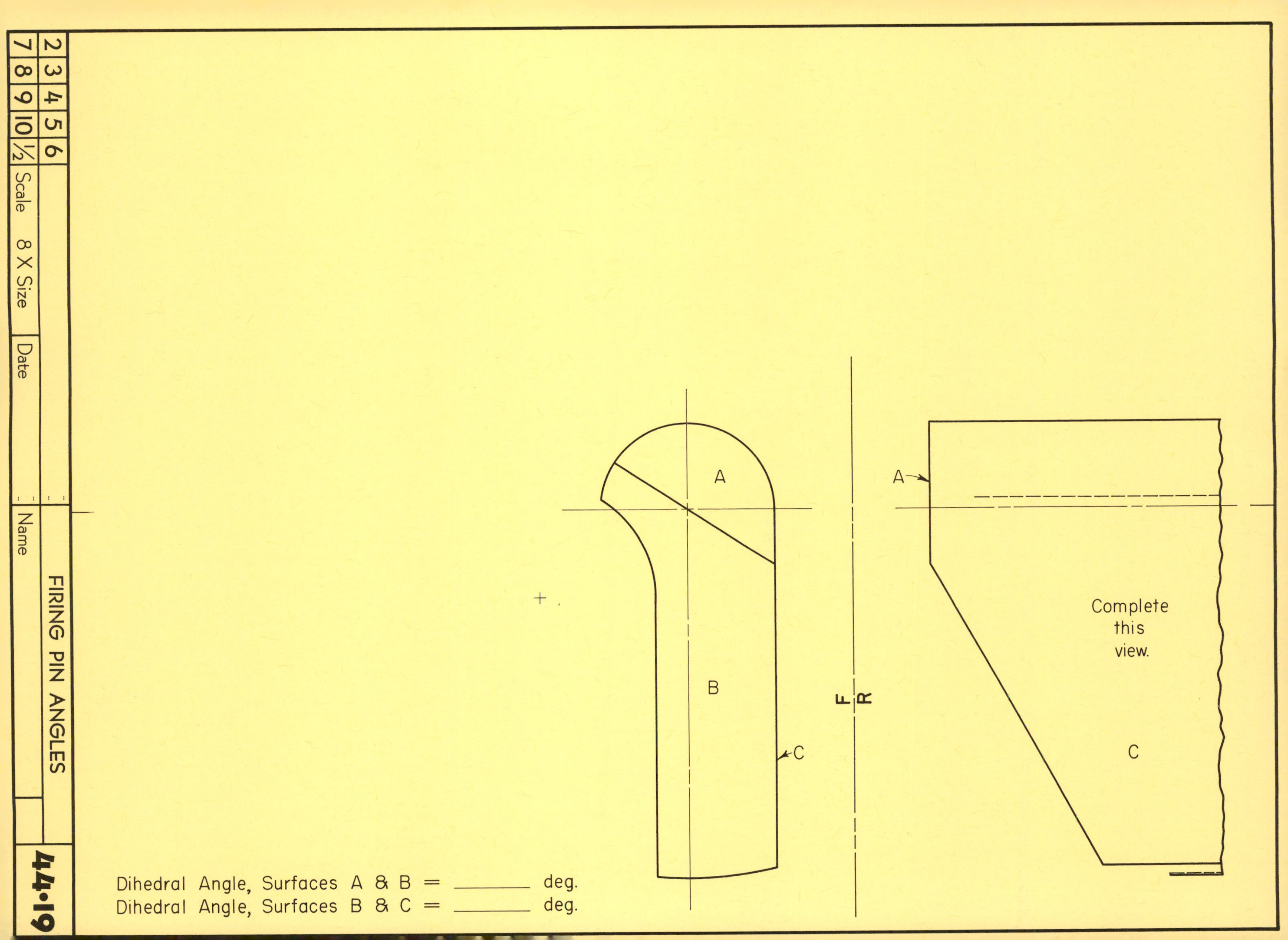
A
A
B
C
F
R
Complete
this
view.
C
Dihedral Angle, Surfaces A & B = ______ deg.
Dihedral Angle, Surfaces B & C = ______ deg.
2 3 4 5 6
7 8 9 10 ½
Scale 8 X Size
Date
Name
FIRING PIN ANGLES
44•19

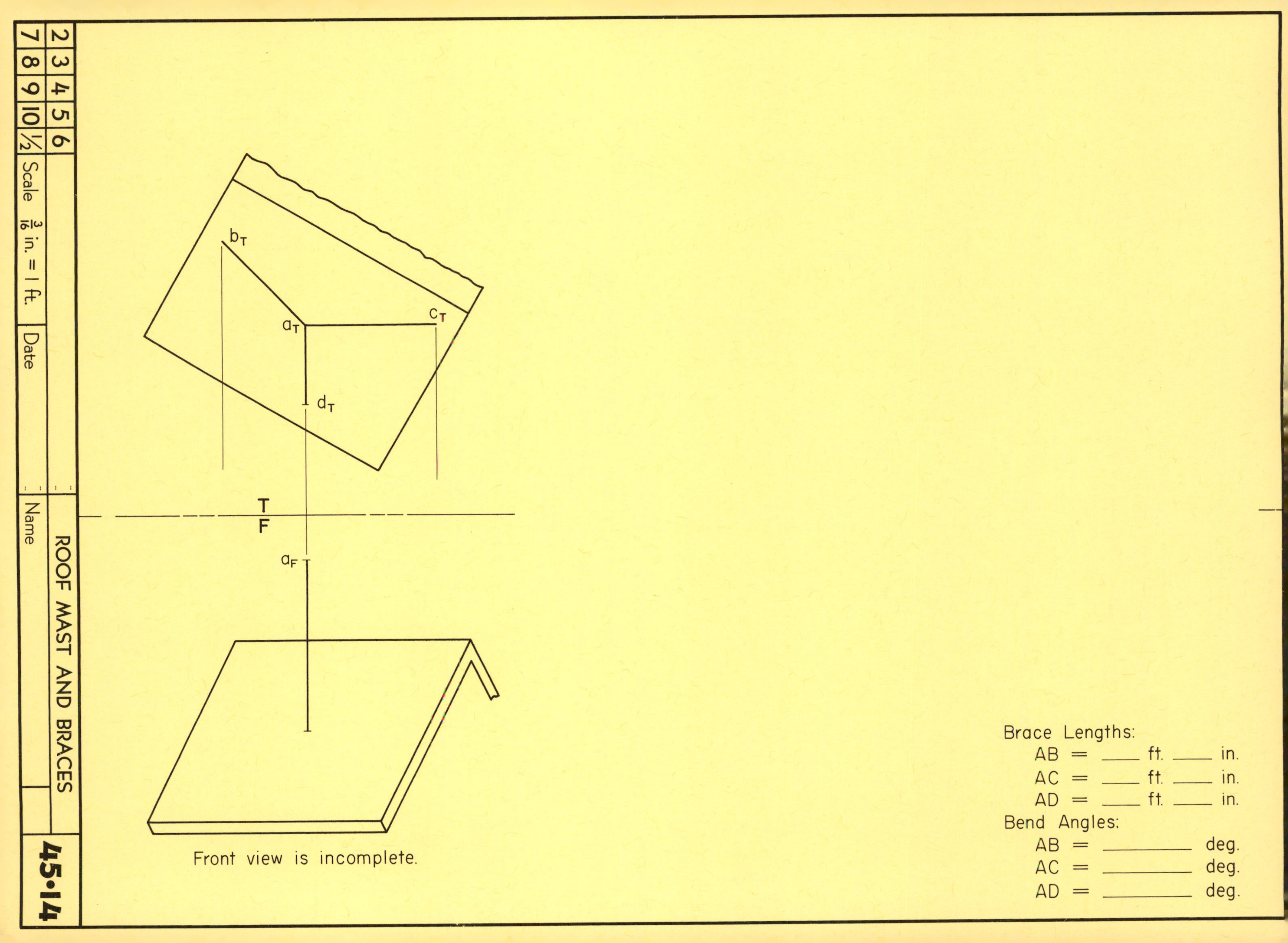

b_T
a_T
c_T
d_T
T
F
a_F
Front view is incomplete.
Brace Lengths:
AB = ____ ft. ____ in.
AC = ____ ft. ____ in.
AD = ____ ft. ____ in.
Bend Angles:
AB = ________ deg.
AC = ________ deg.
AD = ________ deg.
2 3 4 5 6
7 8 9 10 ½
Scale $\frac{3}{16}$ in. = 1 ft.
Date
Name
ROOF MAST AND BRACES
45•14

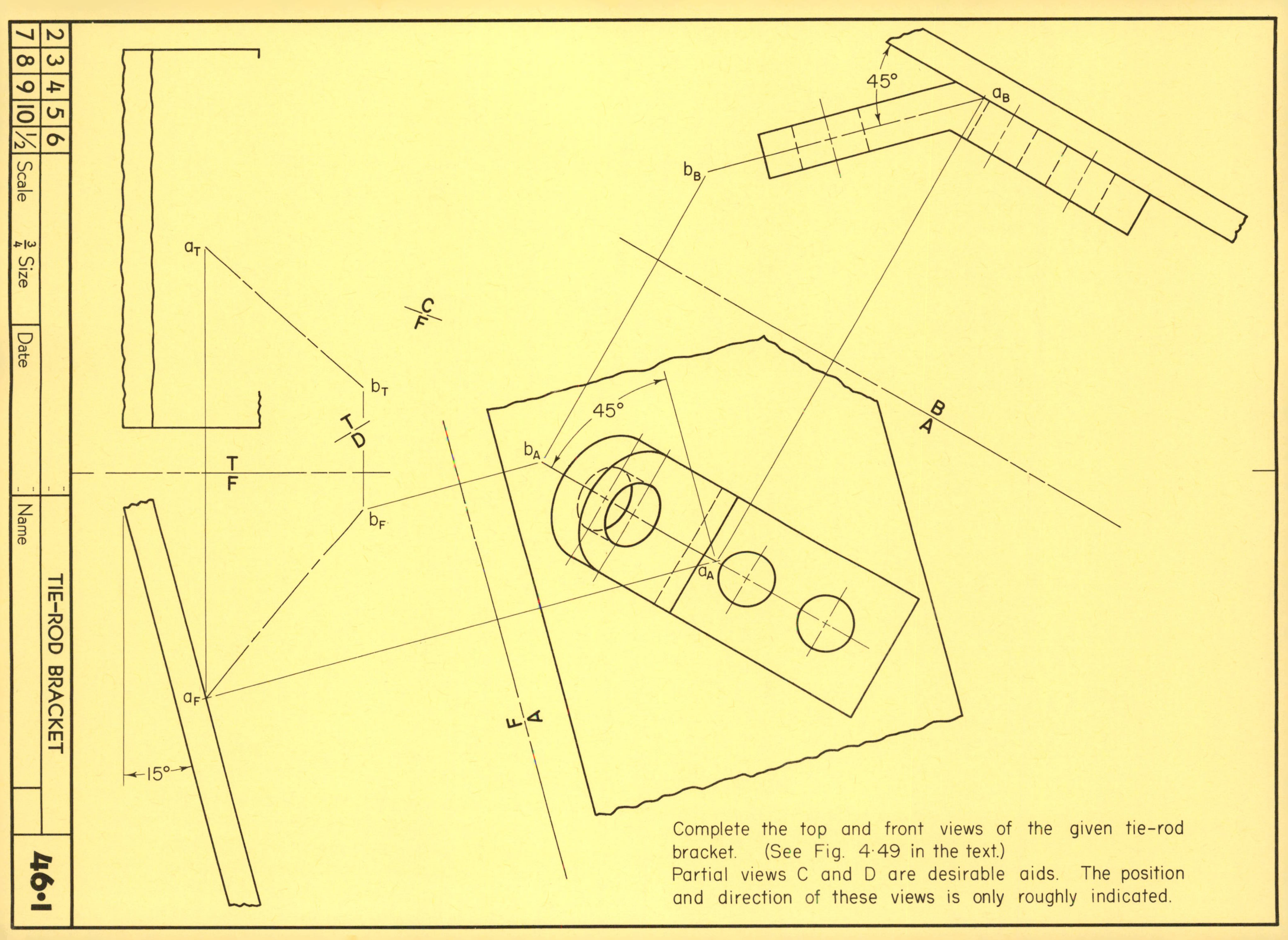

Complete the top and front views of the given tie-rod bracket. (See Fig. 4·49 in the text.)
Partial views C and D are desirable aids. The position and direction of these views is only roughly indicated.

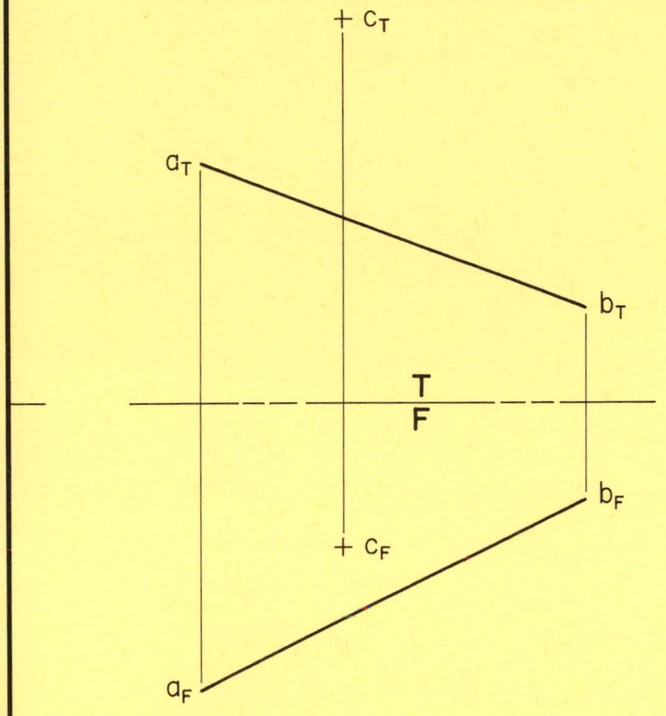

c_T

a_T

b_T

T

F

b_F

c_F

a_F

2	3	4	5	6			REVOLUTION OF A POINT	47•5
7	8	9	10	½	Scale	Date	Name	

Side of Engine — Edge

a_T

b_T

T

F

a_F

Bolt Axis

b_F

Drip Pan — Edge

Wrench can be turned through ______ deg.

Wrench length to clear drip pan = ______ in.

2	3	4	5	6		WRENCH CLEARANCE	47·18
7	8	9	10	½	Scale $1\frac{1}{2}$ in. = 1 ft. Date	Name	

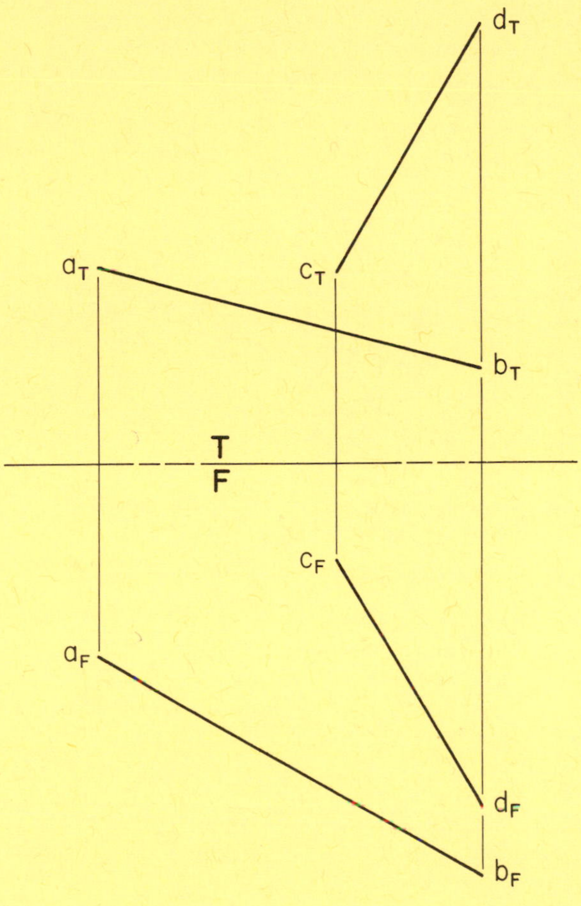

a_T

c_T

d_T

b_T

T

F

c_F

a_F

d_F

b_F

2	3	4	5	6		REVOLUTION OF A LINE	48•7
7	8	9	10	½	Scale	Date	Name	

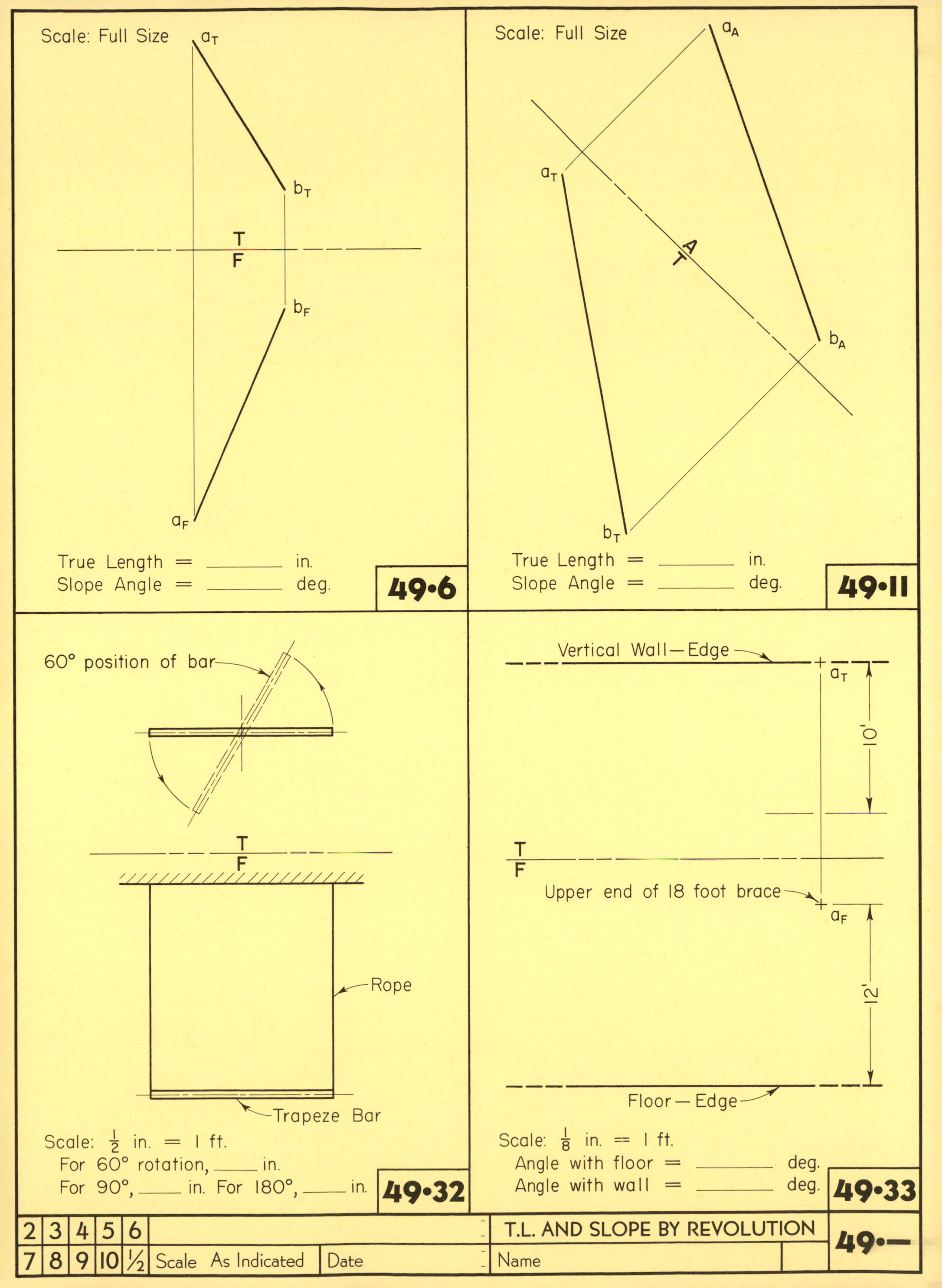

Scale: Full Size
a_T
b_T
T
F
b_F
a_F
True Length = ______ in.
Slope Angle = ______ deg.
49•6
Scale: Full Size
a_A
a_T
A
T
b_A
b_T
True Length = ______ in.
Slope Angle = ______ deg.
49•11
60° position of bar
T
F
Rope
Trapeze Bar
Scale: $\frac{1}{2}$ in. = 1 ft.
For 60° rotation, ____ in.
For 90°, ____ in. For 180°, ____ in.
49•32
Vertical Wall—Edge
a_T
10'
T
F
Upper end of 18 foot brace
a_F
12'
Floor—Edge
Scale: $\frac{1}{8}$ in. = 1 ft.
Angle with floor = ______ deg.
Angle with wall = ______ deg.
49•33
2 3 4 5 6
7 8 9 10 ½
Scale As Indicated
Date
T.L. AND SLOPE BY REVOLUTION
Name
49•—

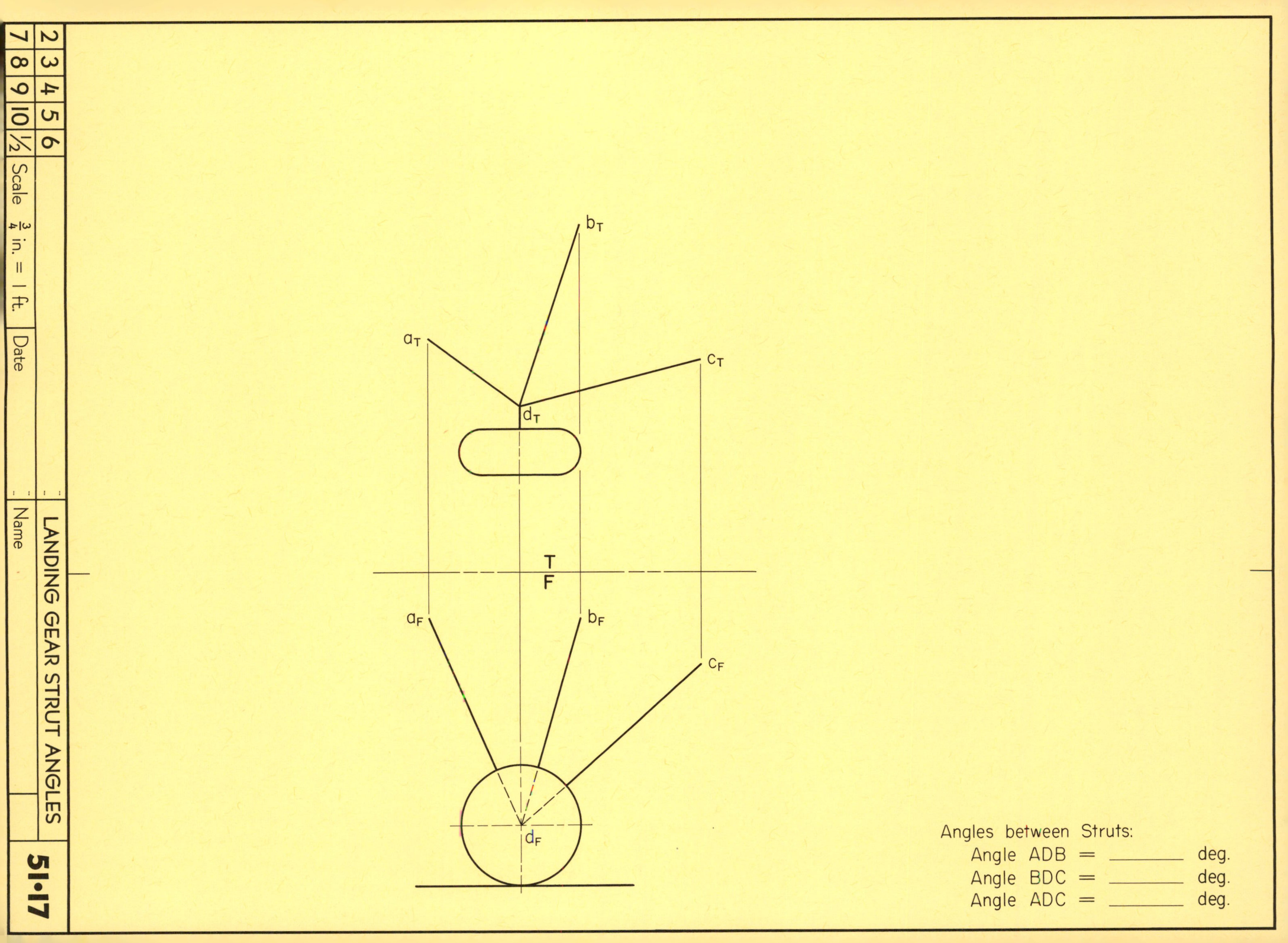

a_T
b_T
c_T
d_T
T
F
a_F
b_F
c_F
d_F
Angles between Struts:
Angle ADB = ______ deg.
Angle BDC = ______ deg.
Angle ADC = ______ deg.
2 3 4 5 6
7 8 9 10 ½
Scale ¾ in. = 1 ft.
Date
LANDING GEAR STRUT ANGLES
Name
51•17

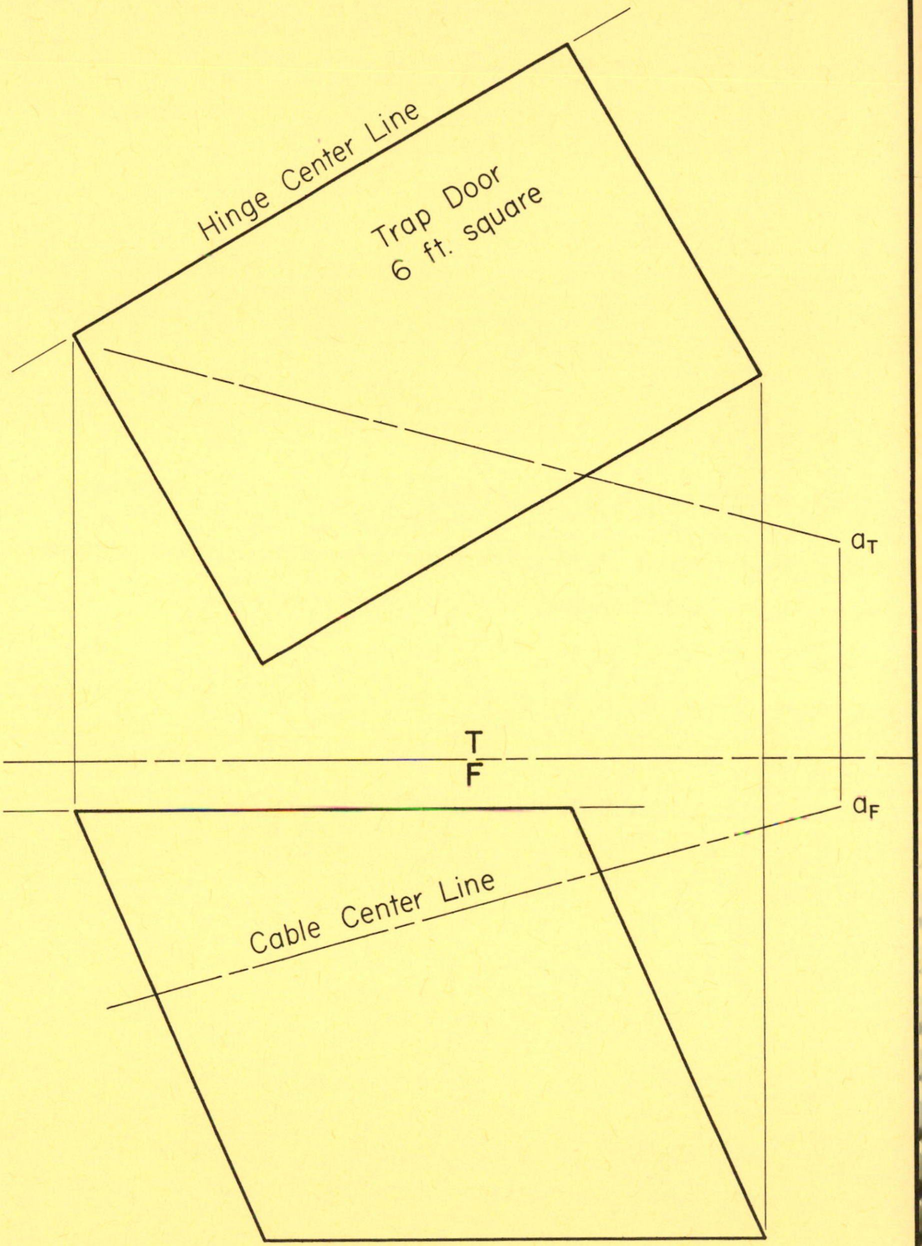

Hinge Center Line

Trap Door
6 ft. square

a_T

T
F

a_F

Cable Center Line

Show the slot center line in the top and front views.
OPTIONAL: Show the slot center line in a revolved true size view, and give offset measurements of the slot from the lower door edge at 6 inch intervals.

2	3	4	5	6		CABLE SLOT IN TRAP DOOR	52•17
7	8	9	10	½	Scale $\frac{1}{2}$ in. = 1 ft. Date	Name	

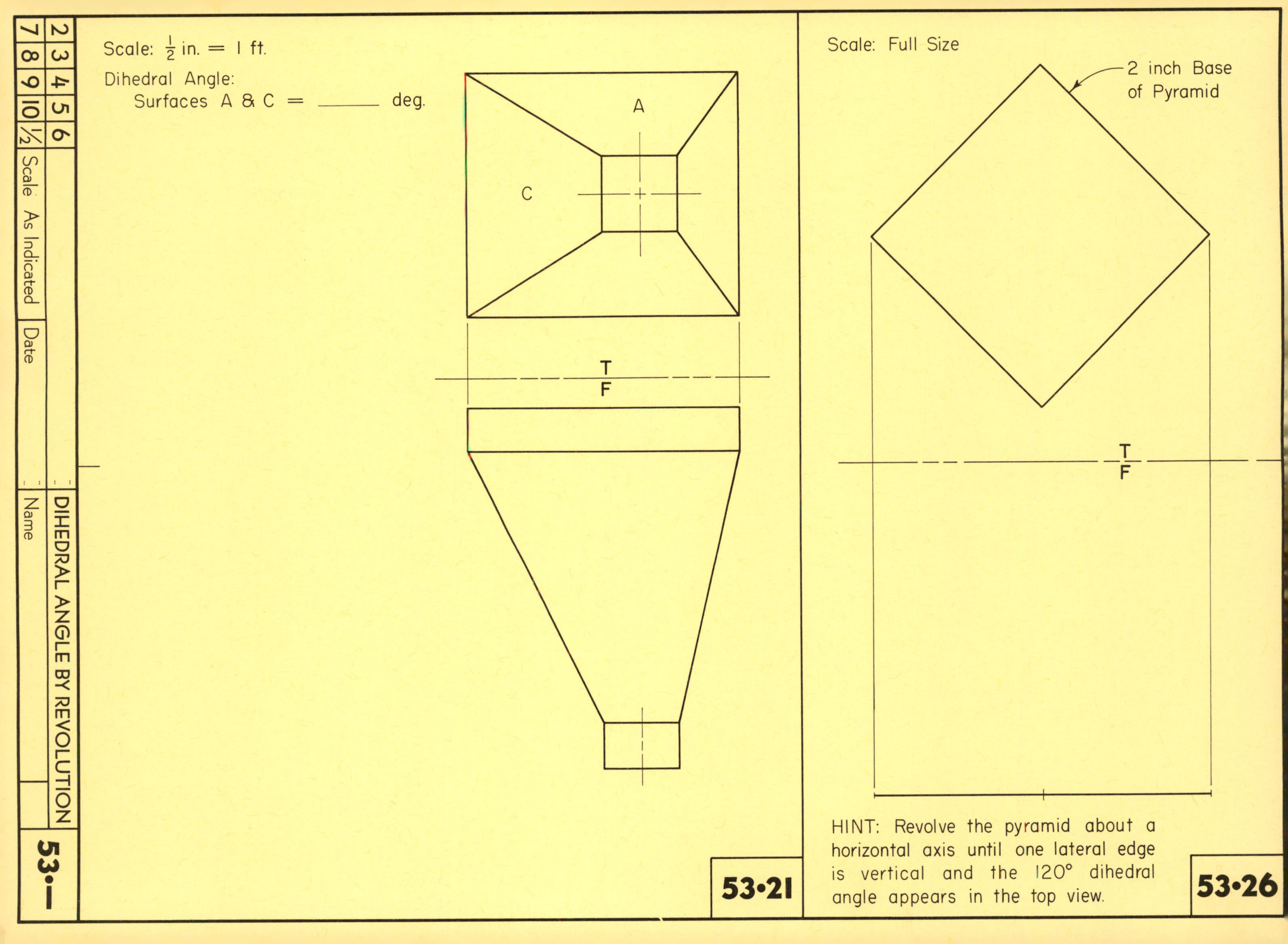
Scale: 1/2 in. = 1 ft.
Dihedral Angle:
Surfaces A & C = ______ deg.
A
C
T
F
53•21
Scale: Full Size
2 inch Base of Pyramid
T
F
HINT: Revolve the pyramid about a horizontal axis until one lateral edge is vertical and the 120° dihedral angle appears in the top view.
53•26
2 3 4 5 6
7 8 9 10 ½
Scale As Indicated
Date
DIHEDRAL ANGLE BY REVOLUTION
Name
53•—

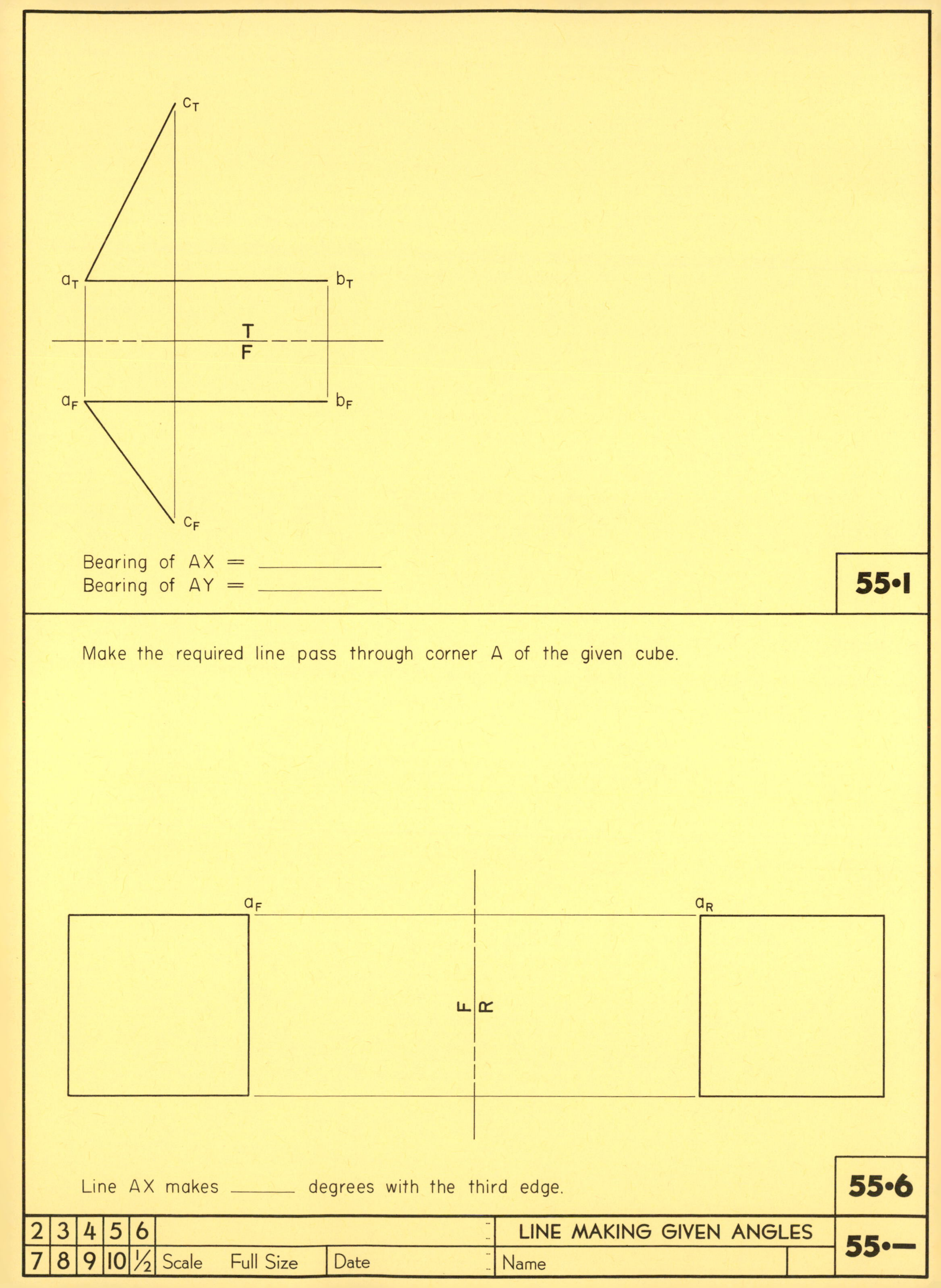

c_T
a_T
b_T
T
F
a_F
b_F
c_F
Bearing of AX =
Bearing of AY =
55•1
Make the required line pass through corner A of the given cube.
a_F
a_R
F
R
Line AX makes degrees with the third edge.
55•6
2 3 4 5 6
7 8 9 10 ½ Scale Full Size Date
LINE MAKING GIVEN ANGLES
Name
55•—

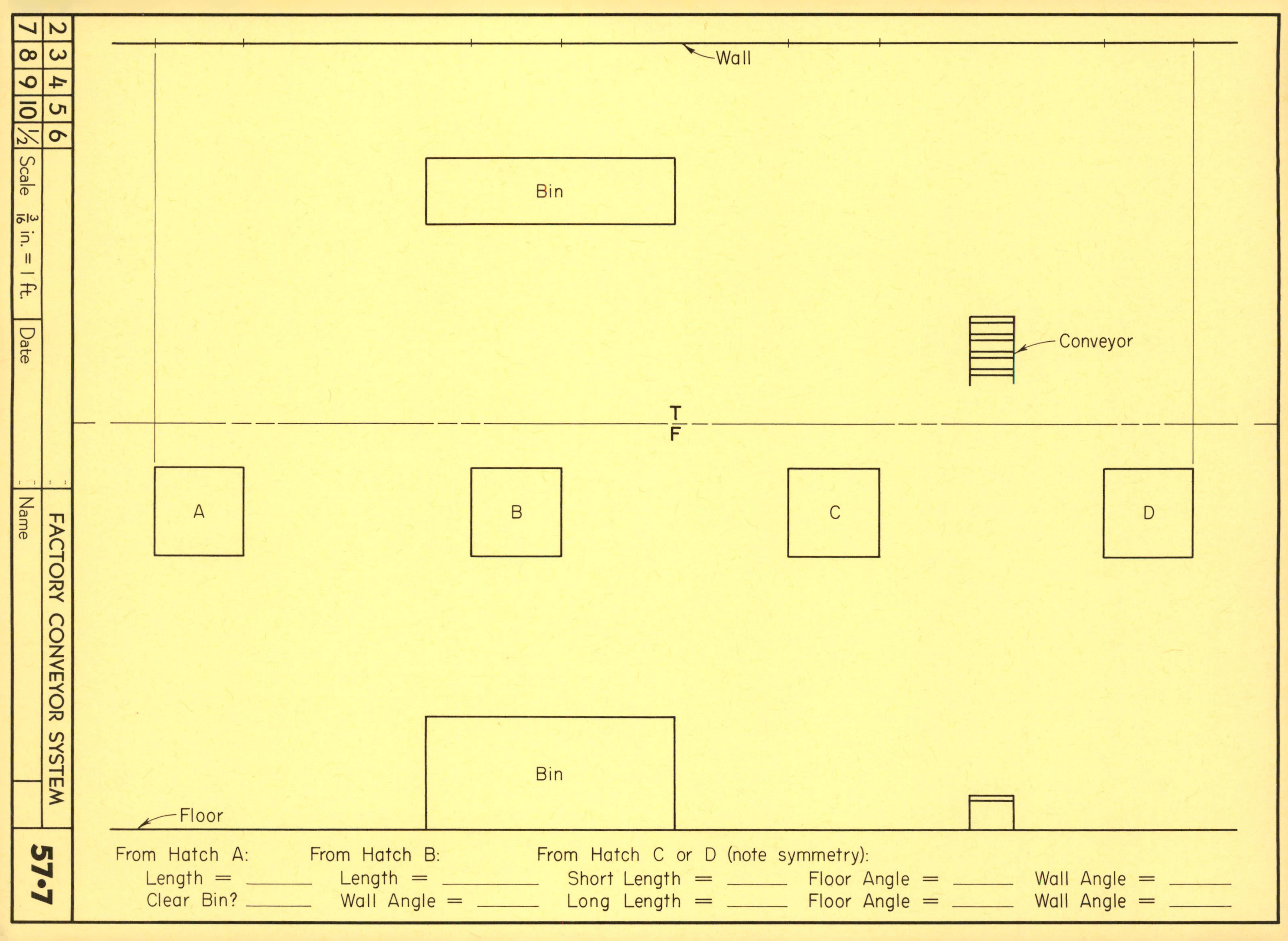

Wall
Bin
Conveyor
T
F
A
B
C
D
Bin
Floor
From Hatch A:
Length = ____
Clear Bin? ____
From Hatch B:
Length = ____
Wall Angle = ____
From Hatch C or D (note symmetry):
Short Length = ____
Long Length = ____
Floor Angle = ____
Floor Angle = ____
Wall Angle = ____
Wall Angle = ____
2 3 4 5 6
7 8 9 10 1/2
Scale 3/16 in. = 1 ft.
Date
Name
FACTORY CONVEYOR SYSTEM
57•7

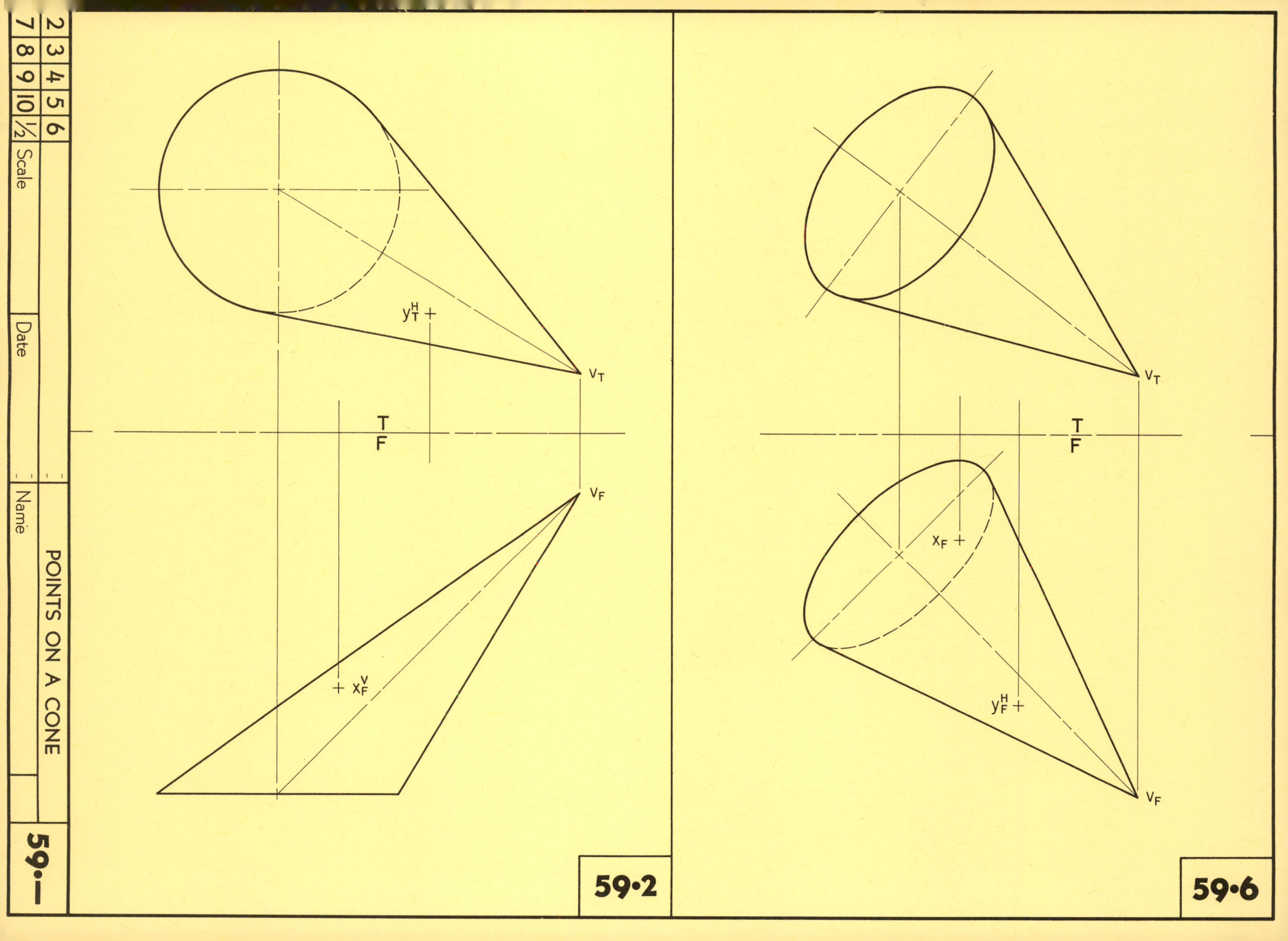

y_T^H
V_T
T
F
V_F
x_F^V
59·2
V_T
T
F
x_F
y_F^H
V_F
59·6
2 3 4 5 6
7 8 9 10 ½
Scale
Date
Name
POINTS ON A CONE
59·—

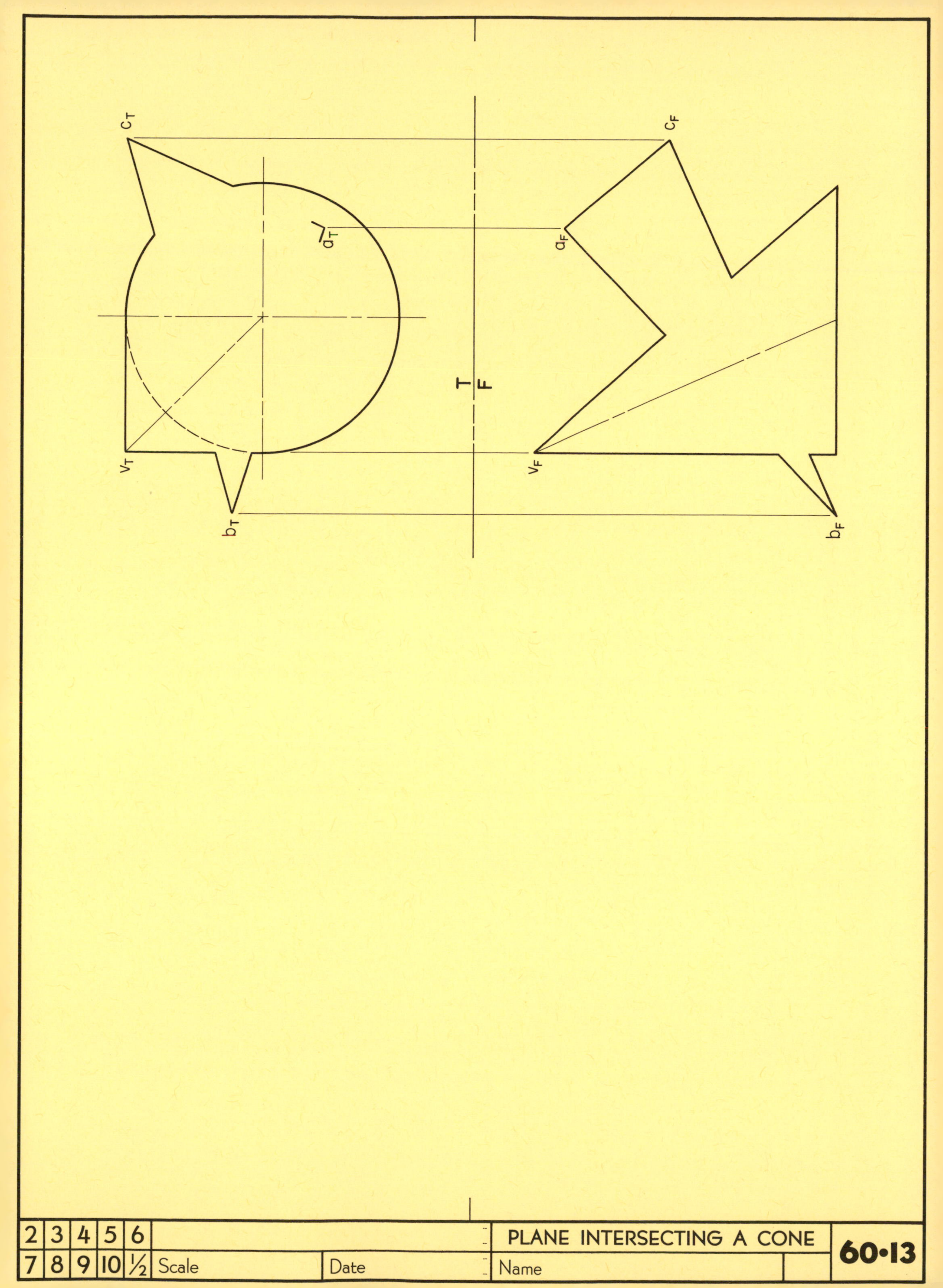

c_T
c_F
a_T
a_F
T
F
v_T
v_F
b_T
b_F
2 3 4 5 6
7 8 9 10 ½
Scale
Date
Name
PLANE INTERSECTING A CONE
60•13

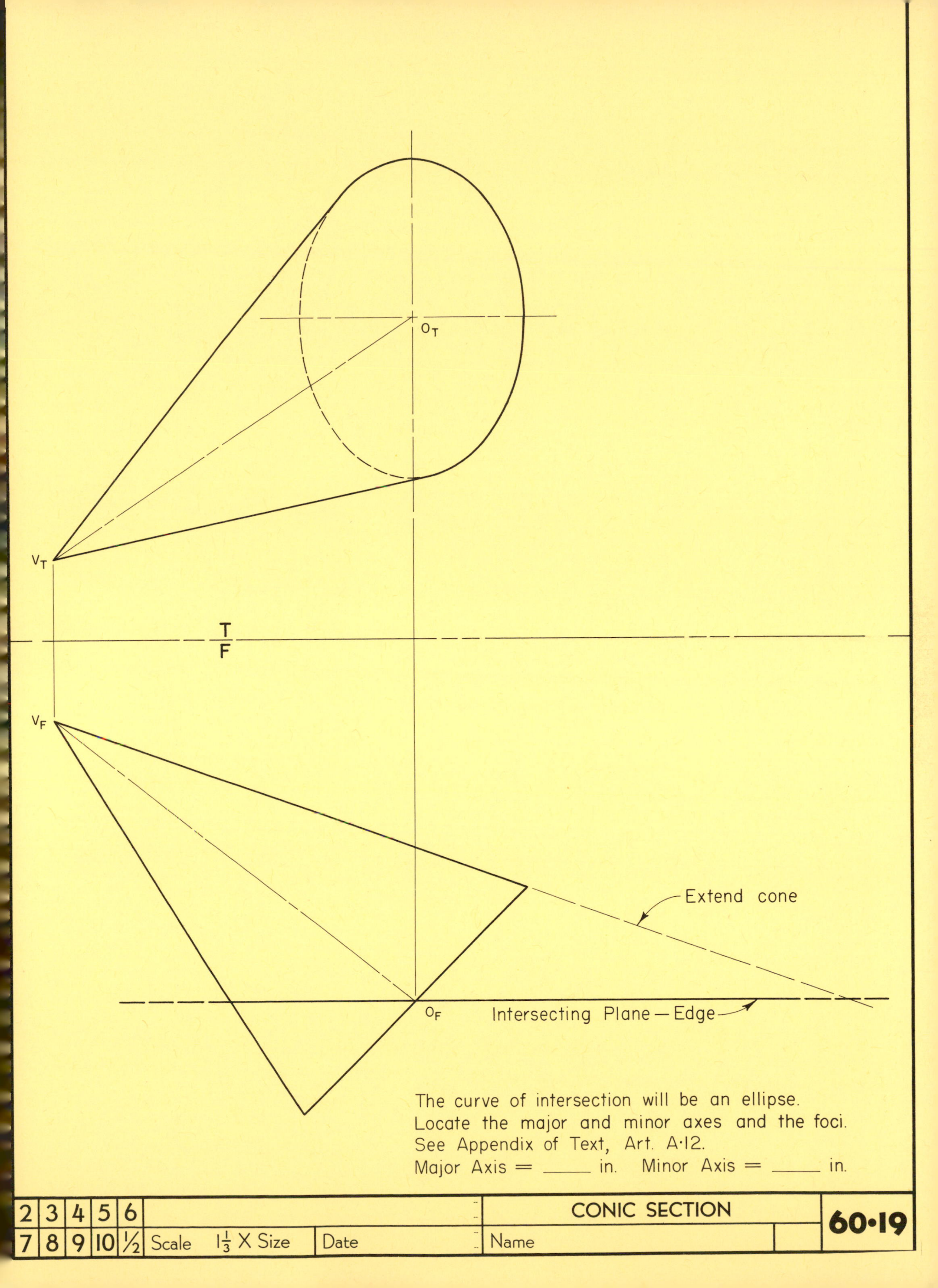
O_T
V_T
T
F
V_F
Extend cone
O_F
Intersecting Plane — Edge
The curve of intersection will be an ellipse.
Locate the major and minor axes and the foci.
See Appendix of Text, Art. A·12.
Major Axis = _____ in. Minor Axis = _____ in.
2 3 4 5 6
7 8 9 10 ½
Scale 1⅓ X Size
Date
CONIC SECTION
Name
60·19

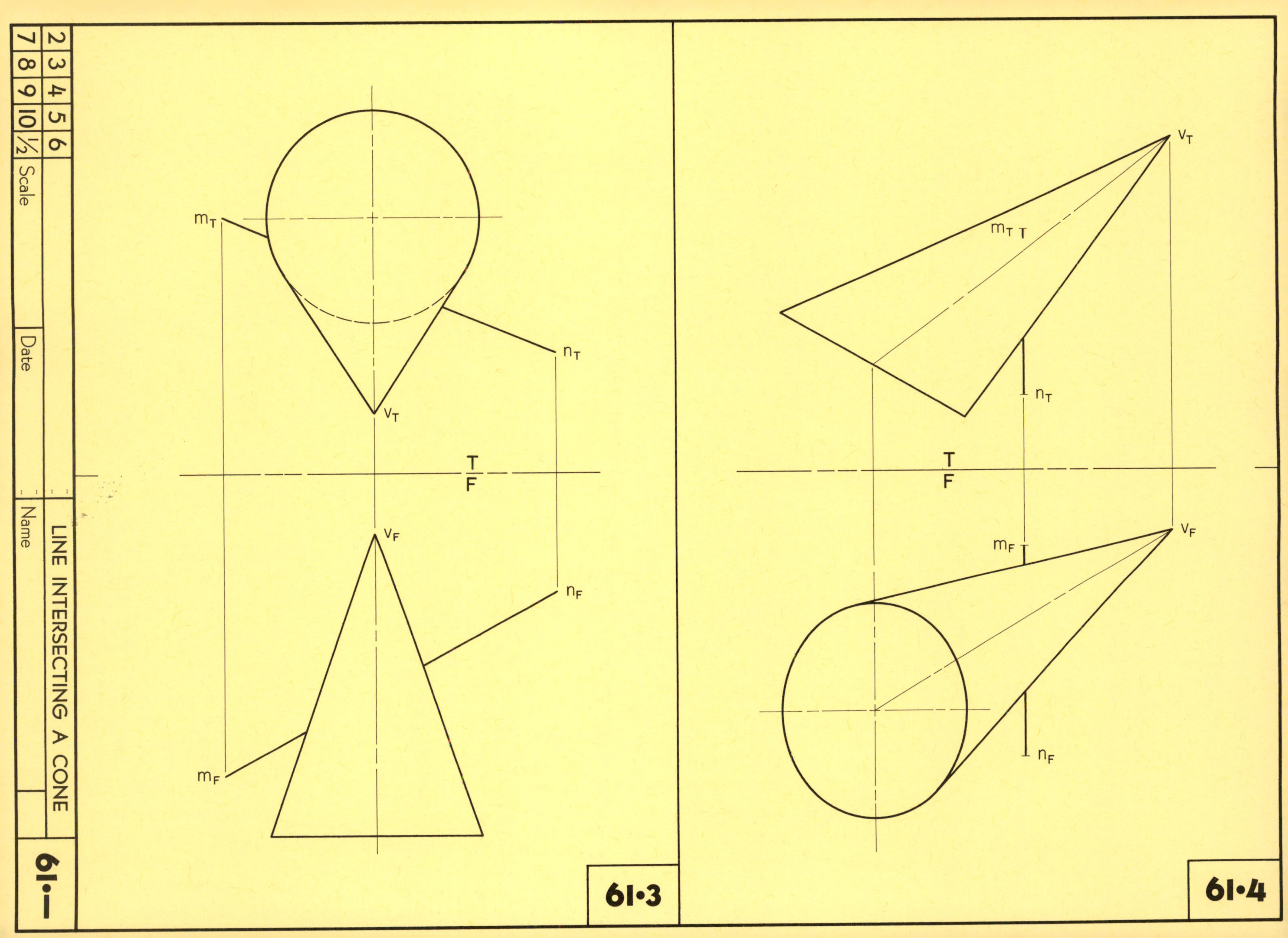
m_T
n_T
v_T
T
F
v_F
n_F
m_F
61·3
v_T
m_T T
n_T
T
F
v_F
m_F
n_F
61·4
2 3 4 5 6
7 8 9 10 ½
Scale
Date
Name
LINE INTERSECTING A CONE
61•—

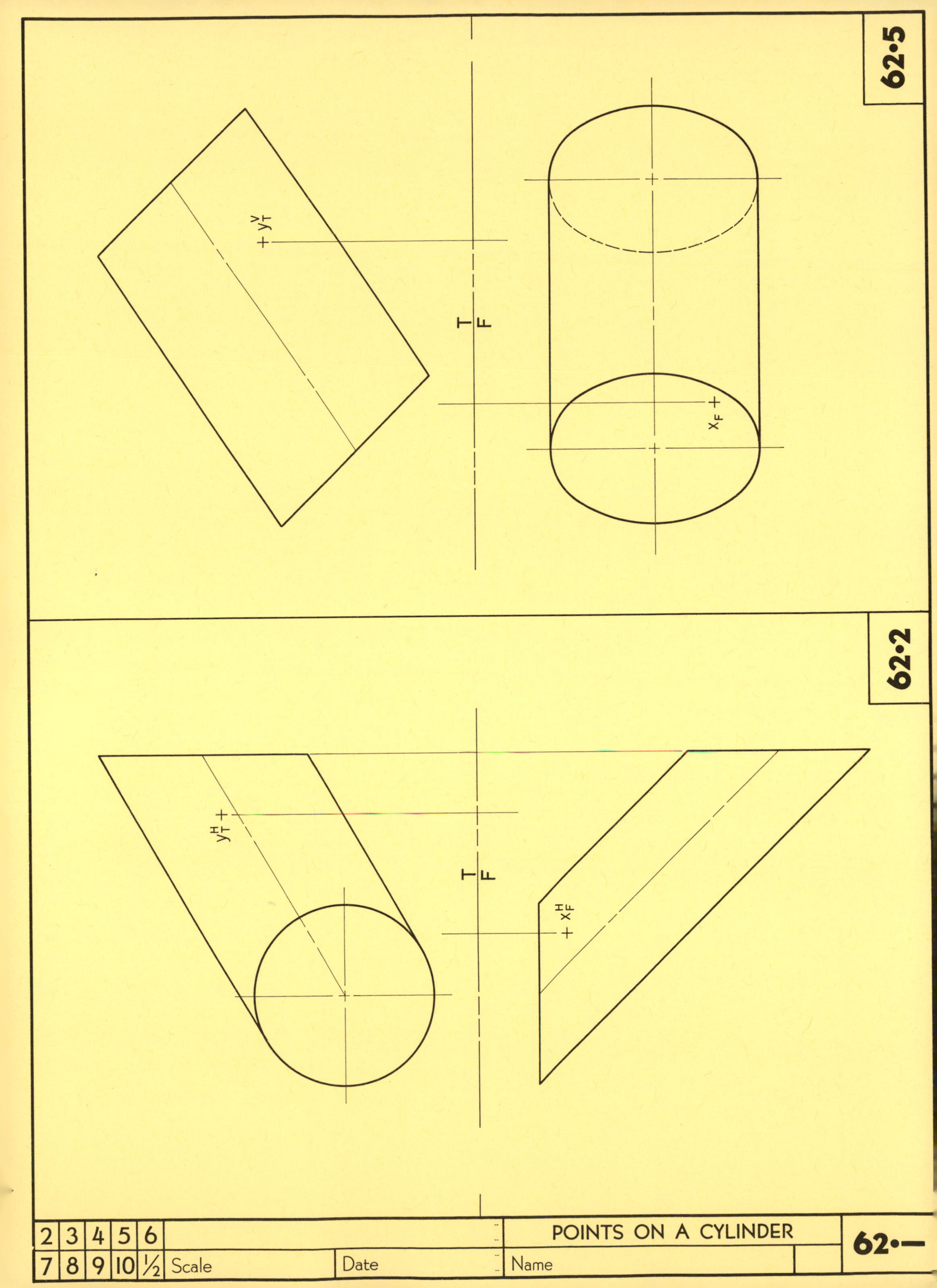
62•5
62•2
T
F
y_T^V
x_F
y_T^H
x_F^H
2 3 4 5 6
7 8 9 10 ½ Scale
Date
Name
POINTS ON A CYLINDER
62•—

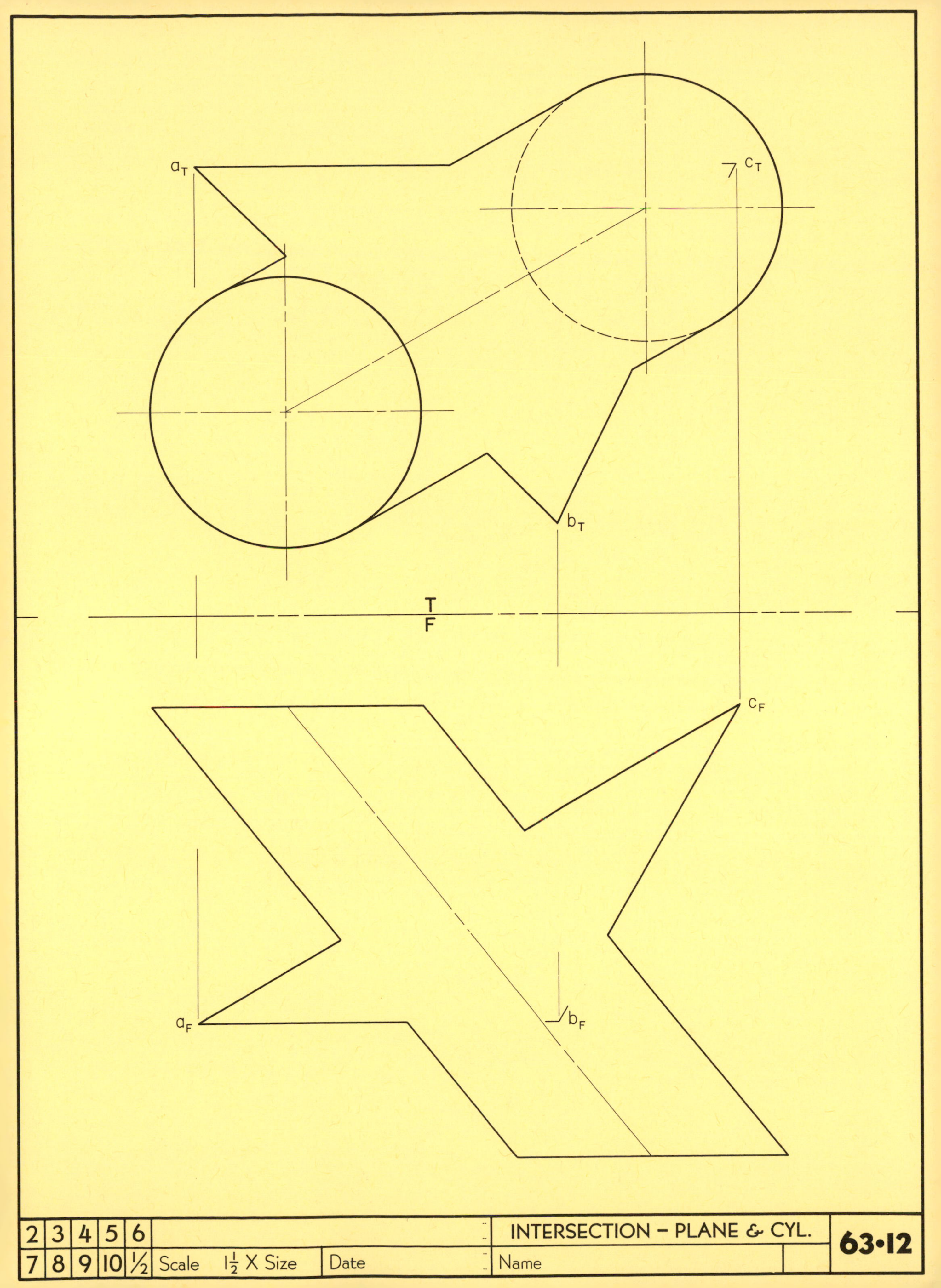

a_T
c_T
b_T
T
F
c_F
a_F
b_F
2 3 4 5 6
7 8 9 10 ½
Scale 1½ X Size
Date
INTERSECTION – PLANE & CYL.
Name
63•12

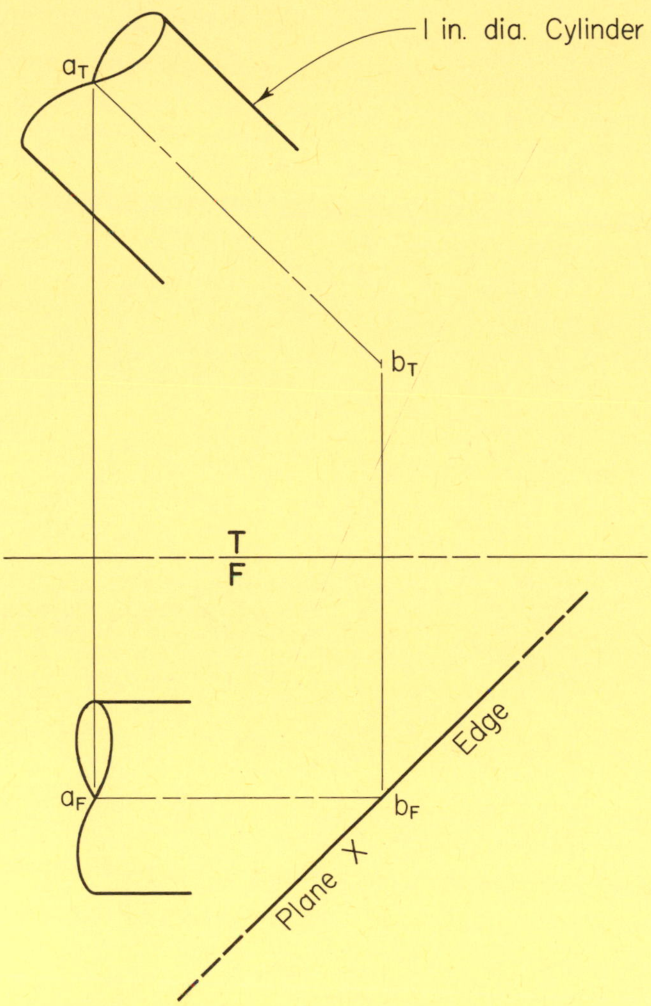

1 in. dia. Cylinder

a_T

b_T

T

F

Edge

a_F

b_F

Plane X

See the parallelogram method for multiview ellipses in text Appendix, Art. A·10.

2	3	4	5	6		RIGHT CIRCULAR CYLINDER	64·6
7	8	9	10	½	Scale Full Size / Date	Name	

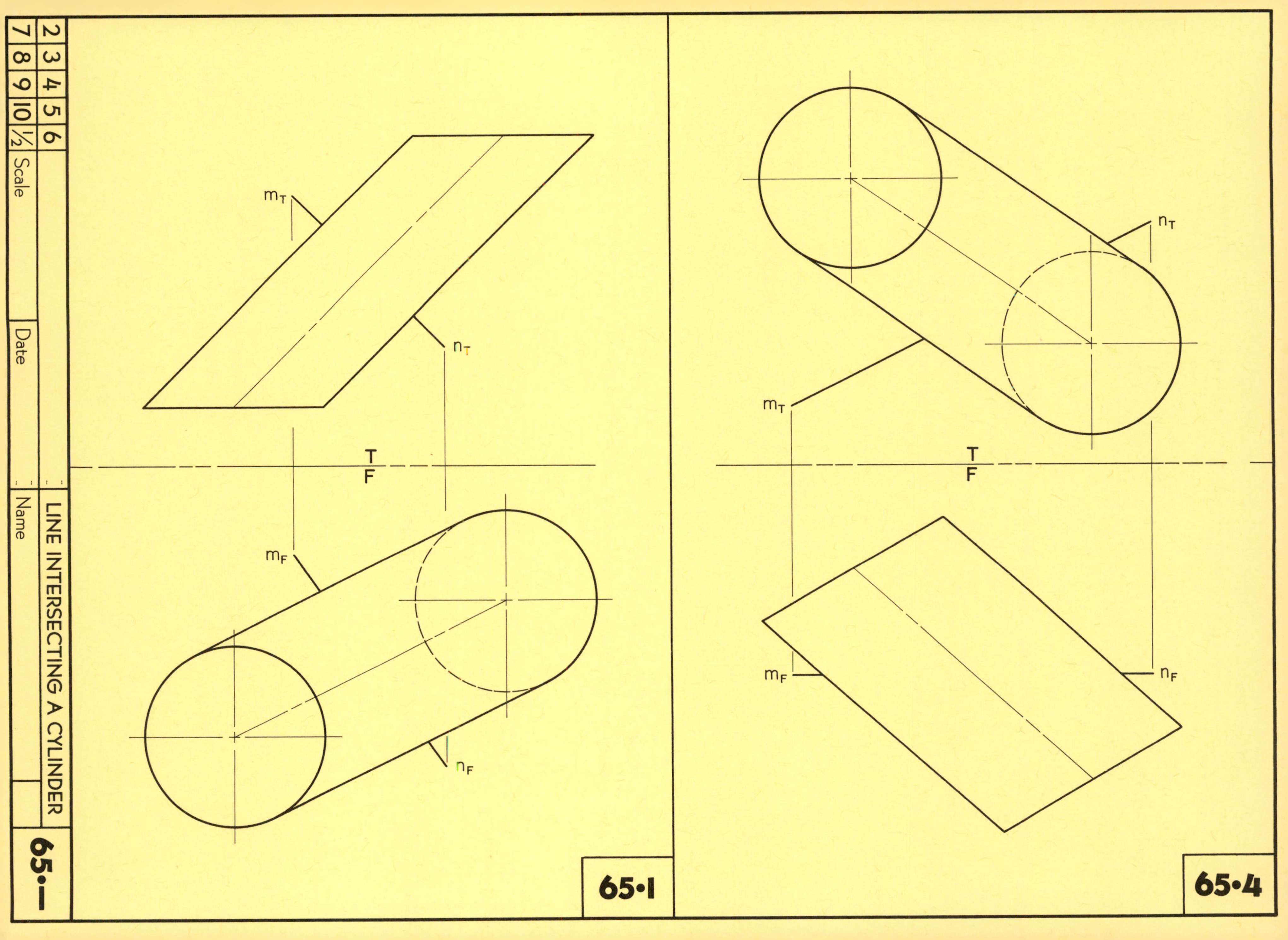

m_T
n_T
T
F
m_F
n_F
65·1
n_T
m_T
T
F
m_F
n_F
65·4
2 3 4 5 6
7 8 9 10 ½
Scale
Date
Name
LINE INTERSECTING A CYLINDER
65·—

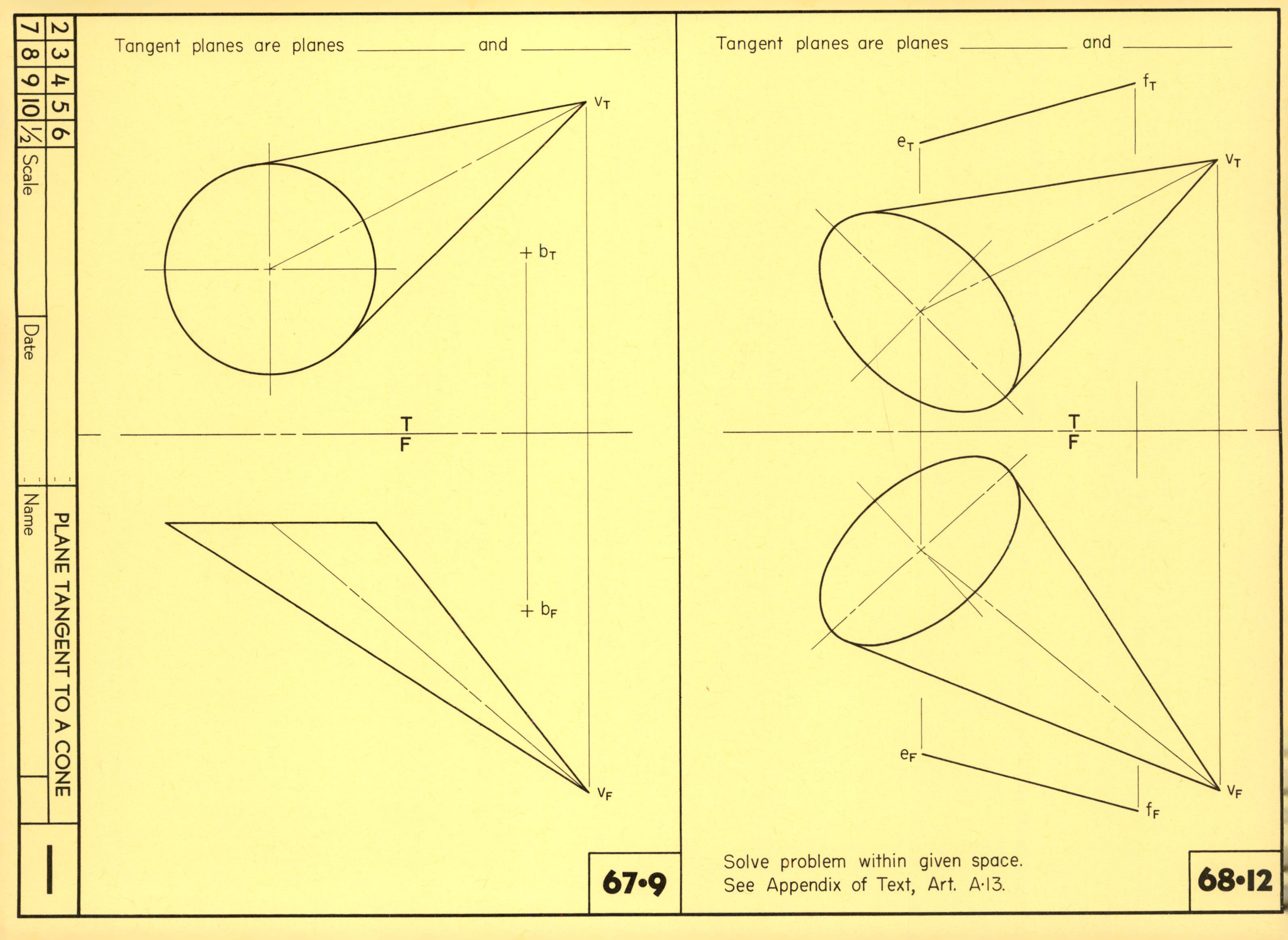

Tangent planes are planes ______ and ______
v_T
b_T
T
F
b_F
v_F
67·9
Tangent planes are planes ______ and ______
f_T
e_T
v_T
T
F
e_F
v_F
f_F
Solve problem within given space.
See Appendix of Text, Art. A·13.
68·12
2 3 4 5 6
7 8 9 10 ½
Scale
Date
Name
PLANE TANGENT TO A CONE

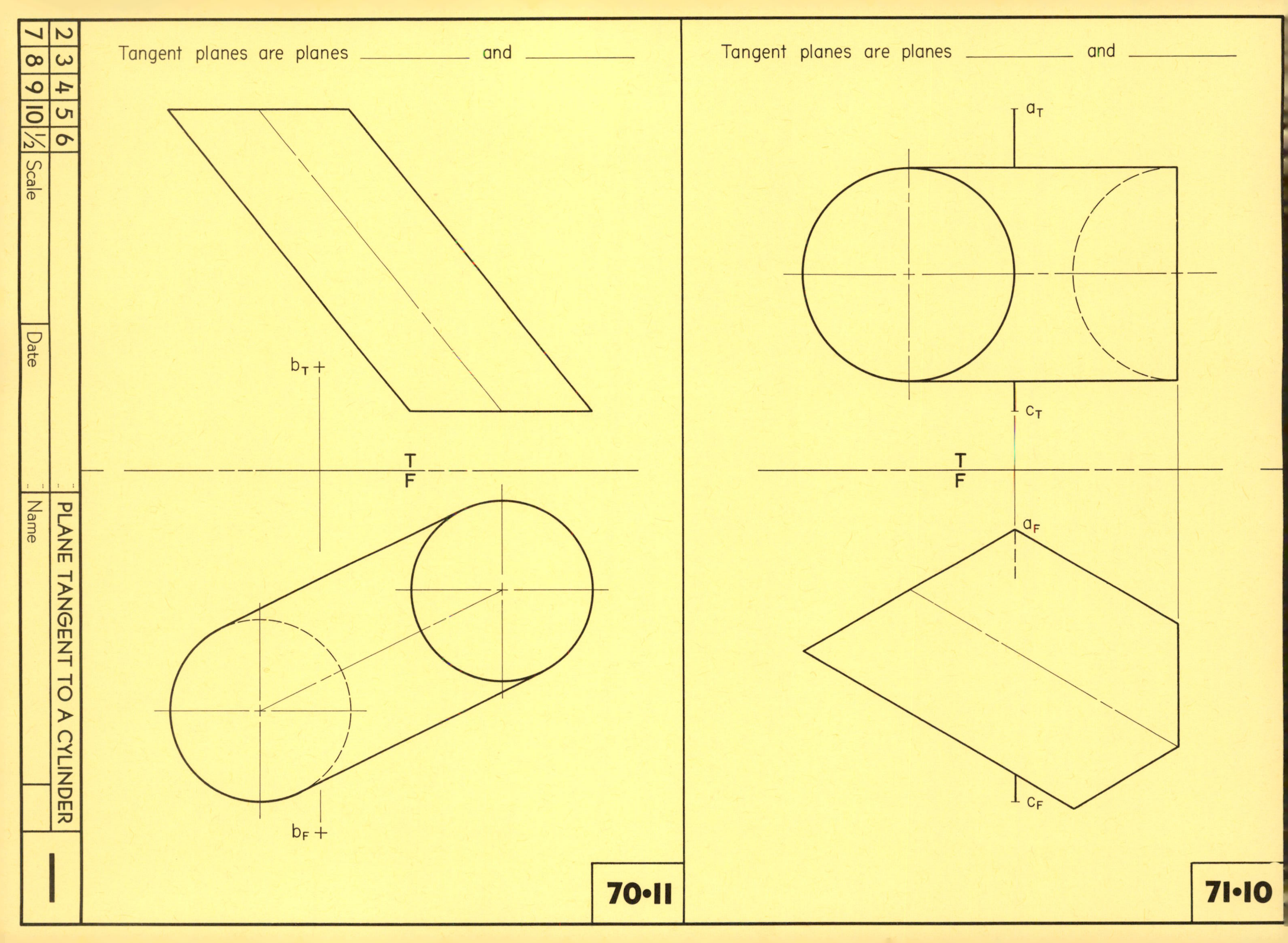
Tangent planes are planes ——— and ———
b_T
T
F
b_F
70·11
Tangent planes are planes ——— and ———
a_T
c_T
T
F
a_F
c_F
71·10
2 3 4 5 6
7 8 9 10 ½
Scale
Date
Name
PLANE TANGENT TO A CYLINDER

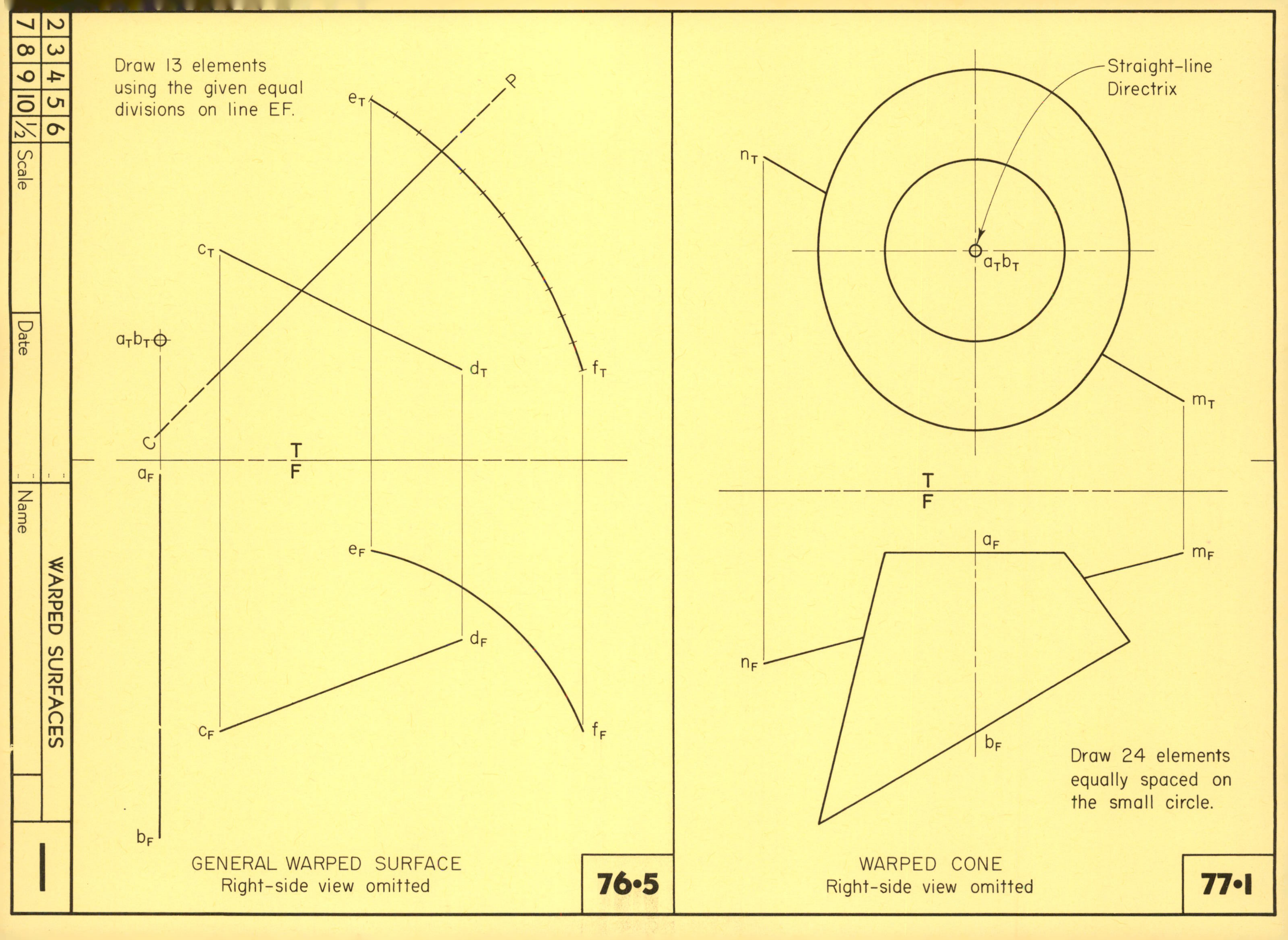

Draw 13 elements using the given equal divisions on line EF.
e_T
P
c_T
a_Tb_T
d_T
f_T
C
T
F
a_F
e_F
d_F
c_F
f_F
b_F
GENERAL WARPED SURFACE
Right-side view omitted
76·5
Straight-line Directrix
n_T
a_Tb_T
m_T
T
F
a_F
m_F
n_F
b_F
Draw 24 elements equally spaced on the small circle.
WARPED CONE
Right-side view omitted
77·1
2 3 4 5 6
7 8 9 10 ½
Scale
Date
Name
WARPED SURFACES

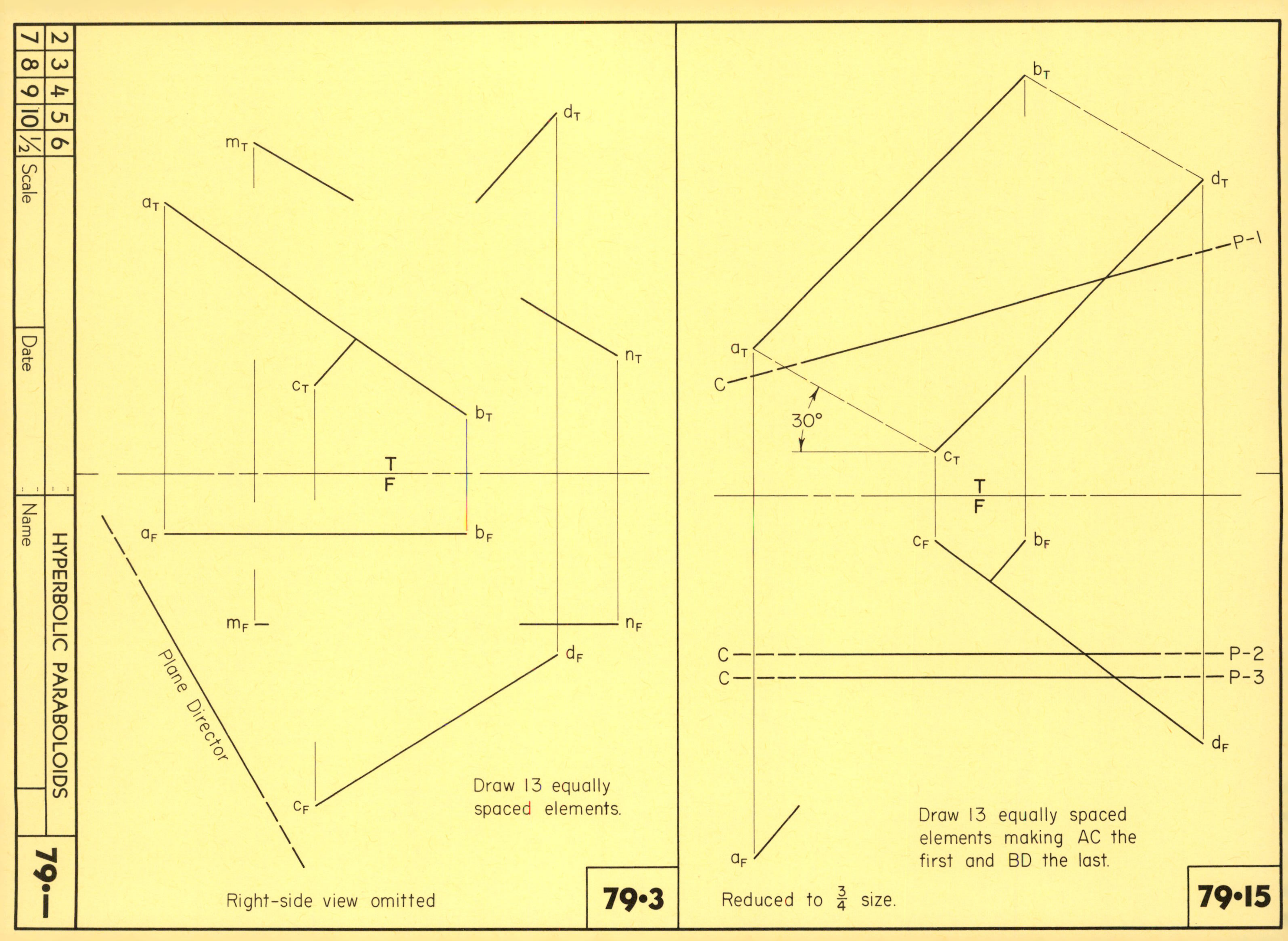

2	3	4	5	6		HYPERBOLIC PARABOLOIDS	79•—
7	8	9	10	½	Scale	Date	Name

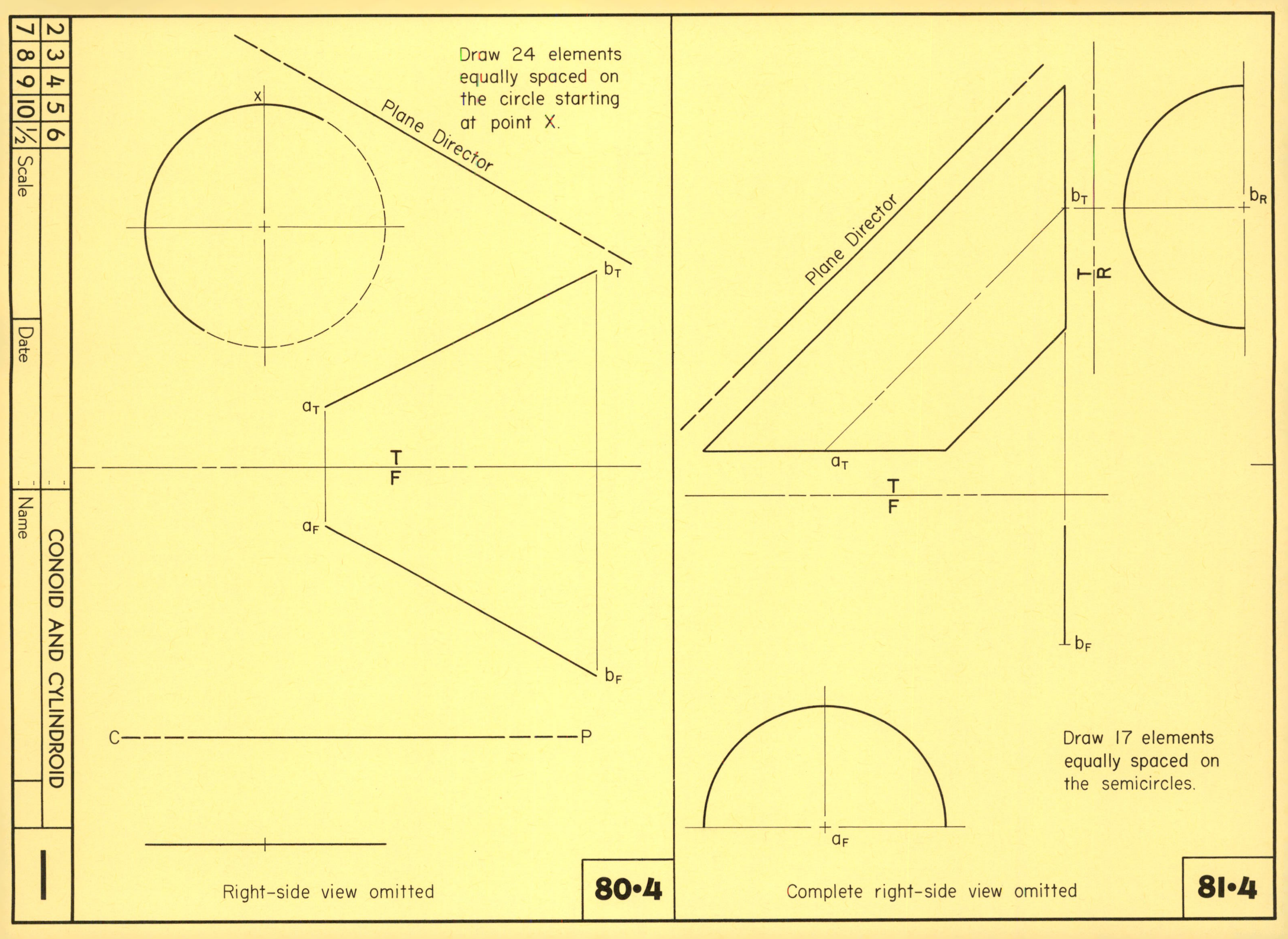
Draw 24 elements equally spaced on the circle starting at point X.
Plane Director
x
a_T
b_T
T
F
a_F
b_F
C
P
Right-side view omitted
80·4
Plane Director
b_T
b_R
T
R
a_T
T
F
b_F
a_F
Draw 17 elements equally spaced on the semicircles.
Complete right-side view omitted
81·4
2 3 4 5 6
7 8 9 10 ½
Scale
Date
Name
CONOID AND CYLINDROID

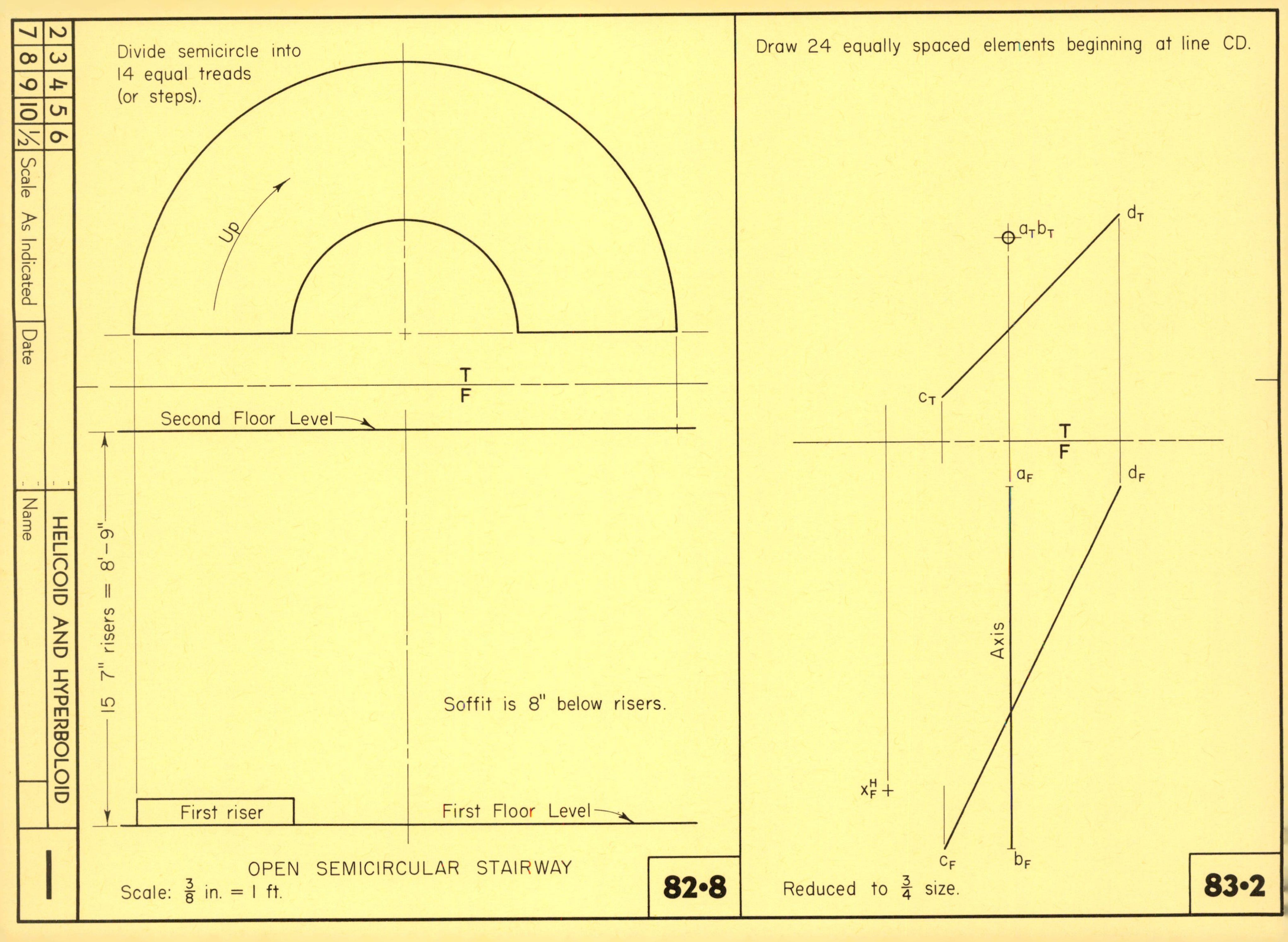

Divide semicircle into
14 equal treads
(or steps).
Up
T
F
Second Floor Level
15 7" risers = 8'-9"
Soffit is 8" below risers.
First riser
First Floor Level
OPEN SEMICIRCULAR STAIRWAY
Scale: 3/8 in. = 1 ft.
82•8
Draw 24 equally spaced elements beginning at line CD.
$a_T b_T$
d_T
c_T
T
F
a_F
d_F
Axis
x_F^H
c_F
b_F
Reduced to 3/4 size.
83•2
2 3 4 5 6
7 8 9 10 ½
Scale As Indicated
Date
Name
HELICOID AND HYPERBOLOID

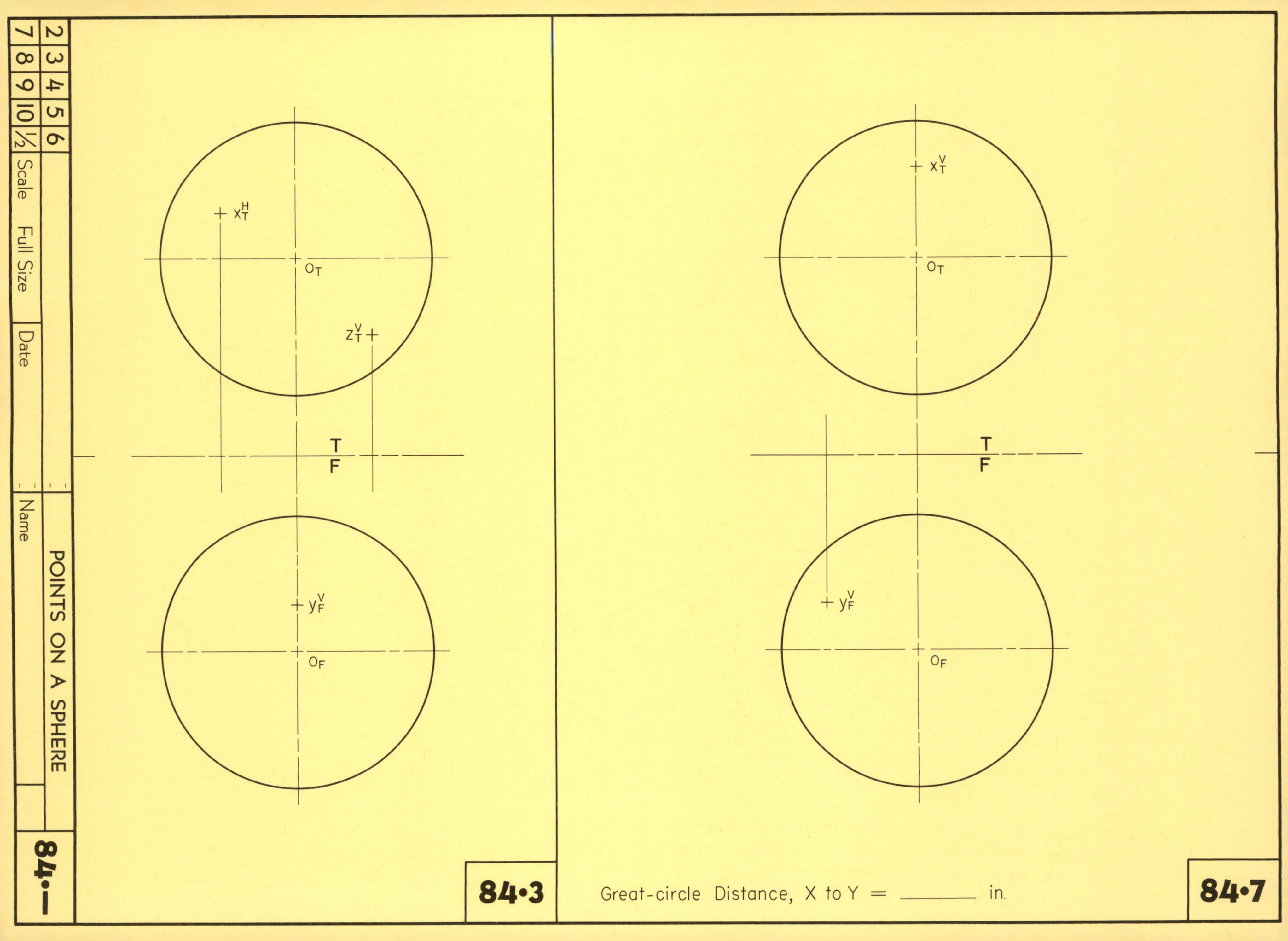
x_T^H
O_T
z_T^V
T
F
y_F^V
O_F
84·3
x_T^V
O_T
T
F
y_F^V
O_F
Great-circle Distance, X to Y = ______ in.
84·7
2 3 4 5 6
7 8 9 10 ½
Scale Full Size
Date
Name
POINTS ON A SPHERE
84·—

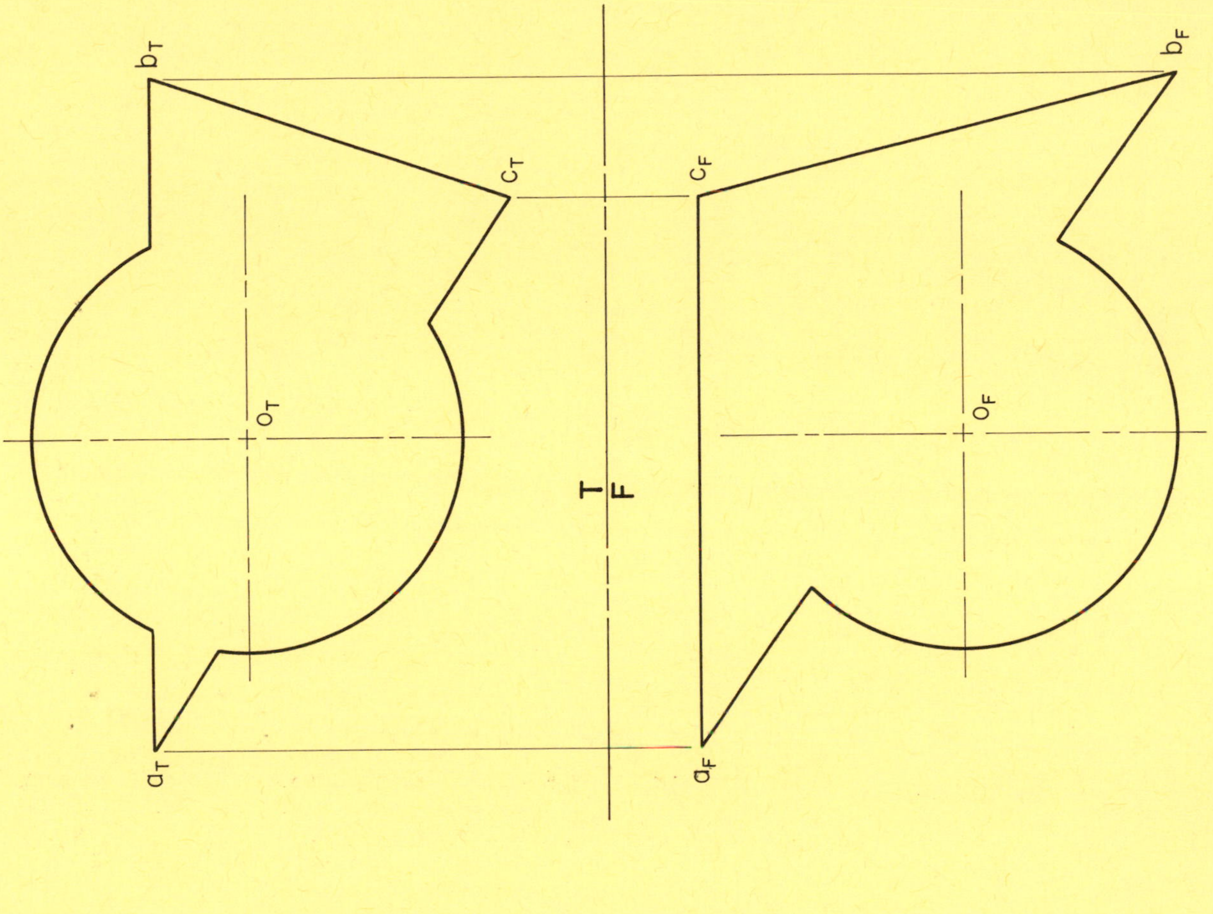

b_T c_T o_T a_T

T
F

b_F c_F o_F a_F

2	3	4	5	6		PLANE INTERSECTING SPHERE	85•8
7	8	9	10	½	Scale	Date	Name

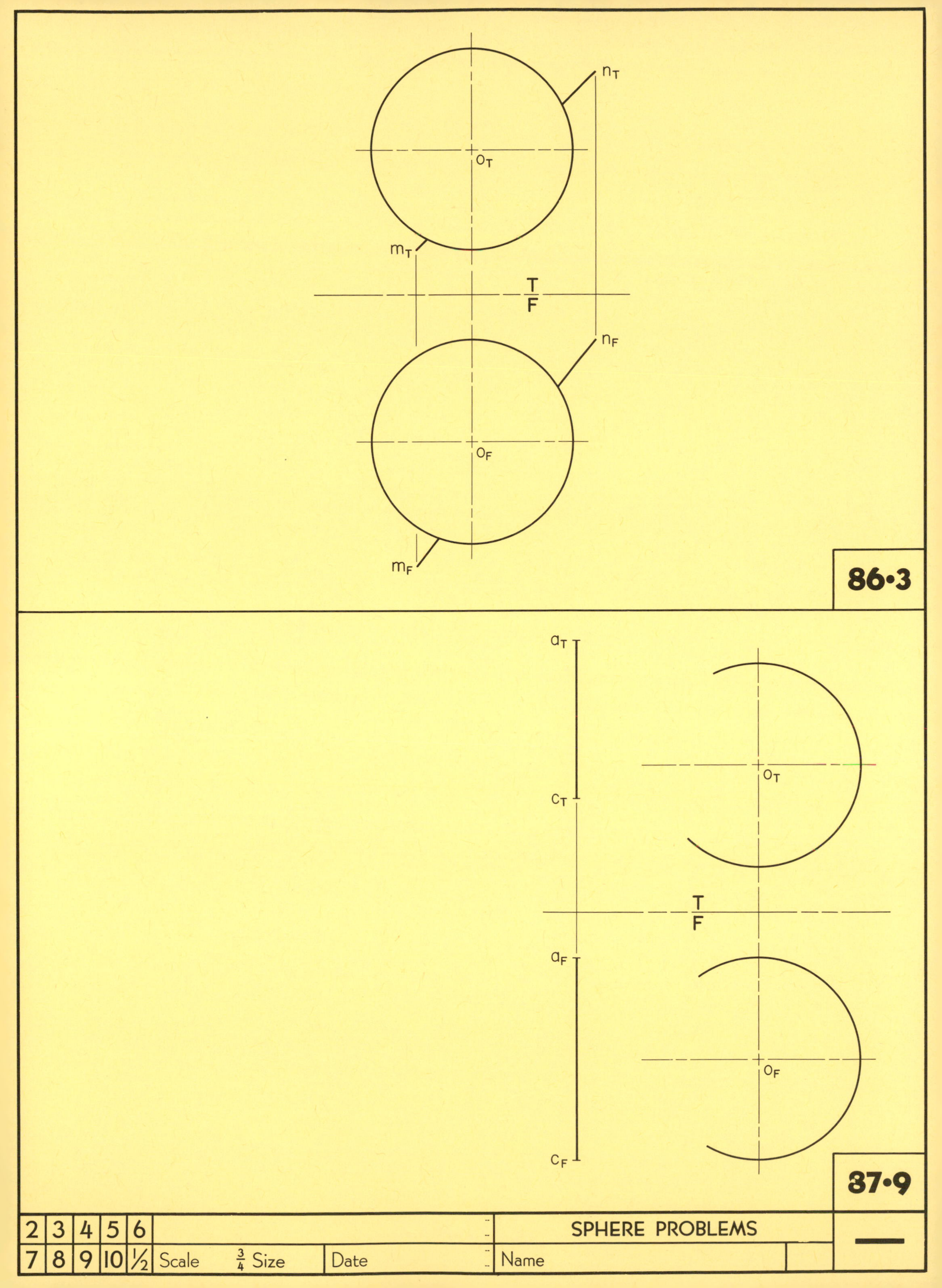

n_T
o_T
m_T
T
F
n_F
o_F
m_F
86•3
a_T
o_T
c_T
T
F
a_F
o_F
c_F
87•9
2 3 4 5 6
7 8 9 10 ½
Scale 3/4 Size
Date
SPHERE PROBLEMS
Name

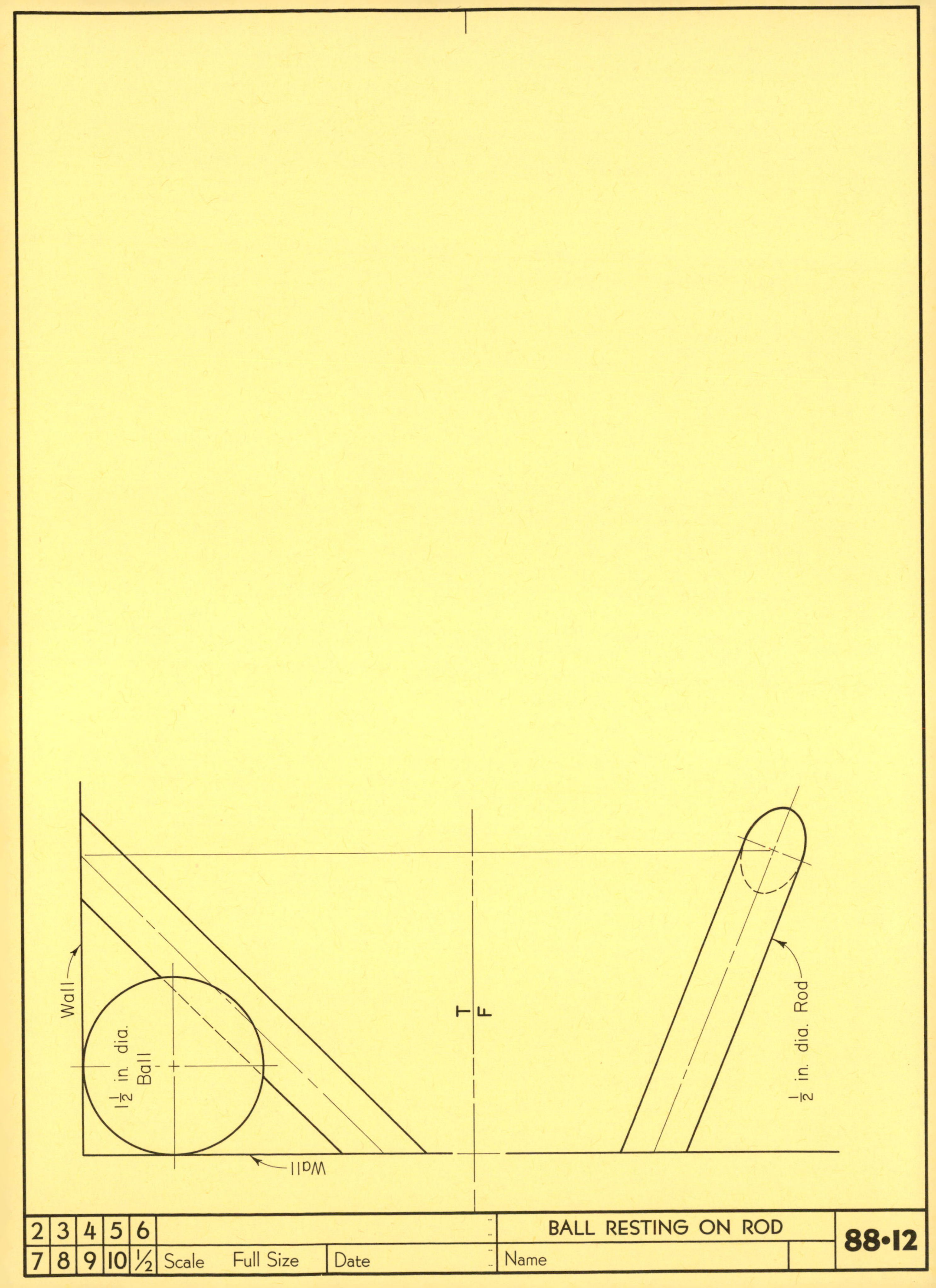
Wall
1½ in. dia. Ball
Wall
T
F
½ in. dia. Rod
2 3 4 5 6
7 8 9 10 ½
Scale Full Size
Date
BALL RESTING ON ROD
Name
88•12

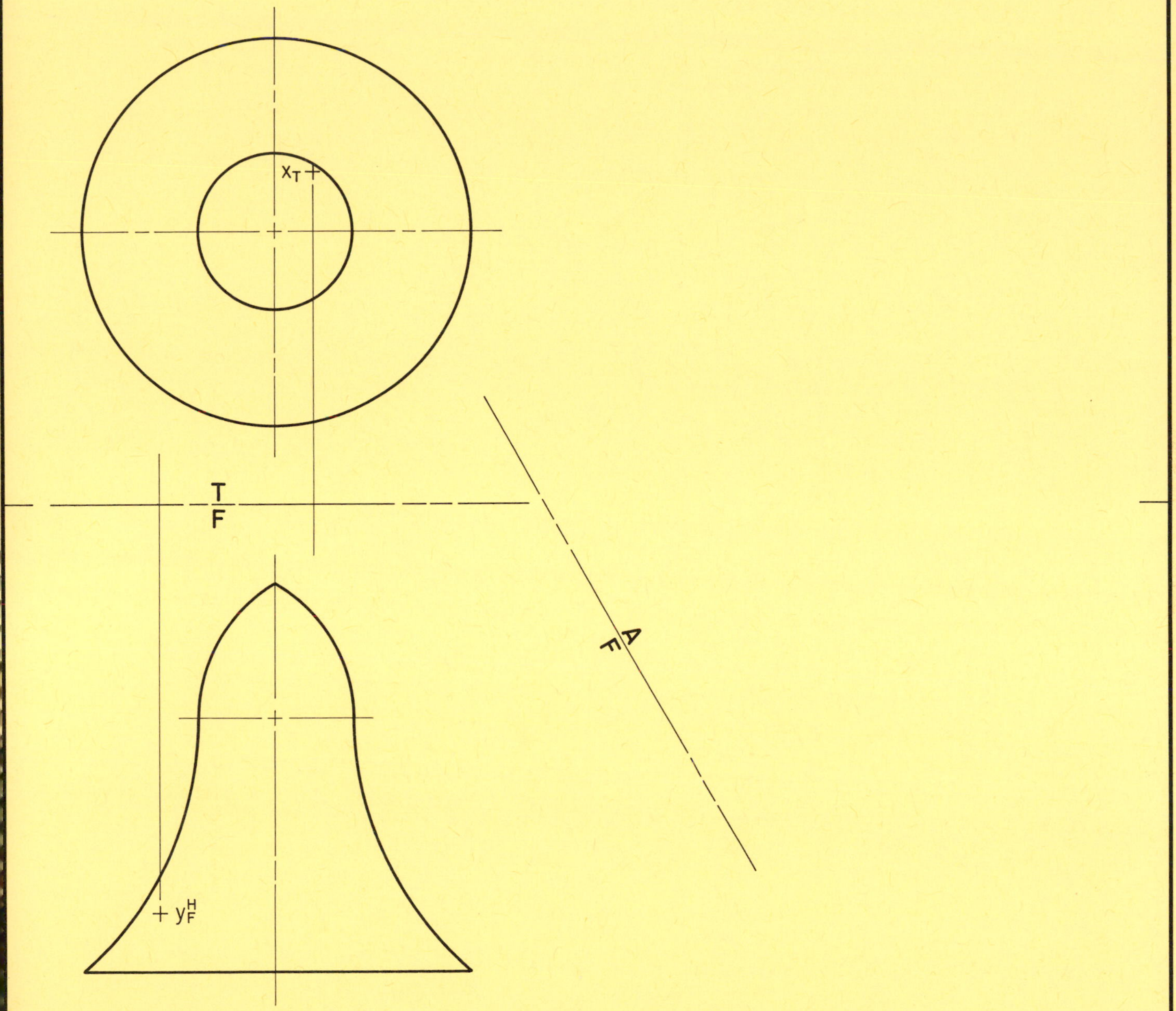

x_T

T

F

A

F

y_F^H

2	3	4	5	6			SURFACE OF REVOLUTION	89·4
7	8	9	10	½	Scale	Date	Name	

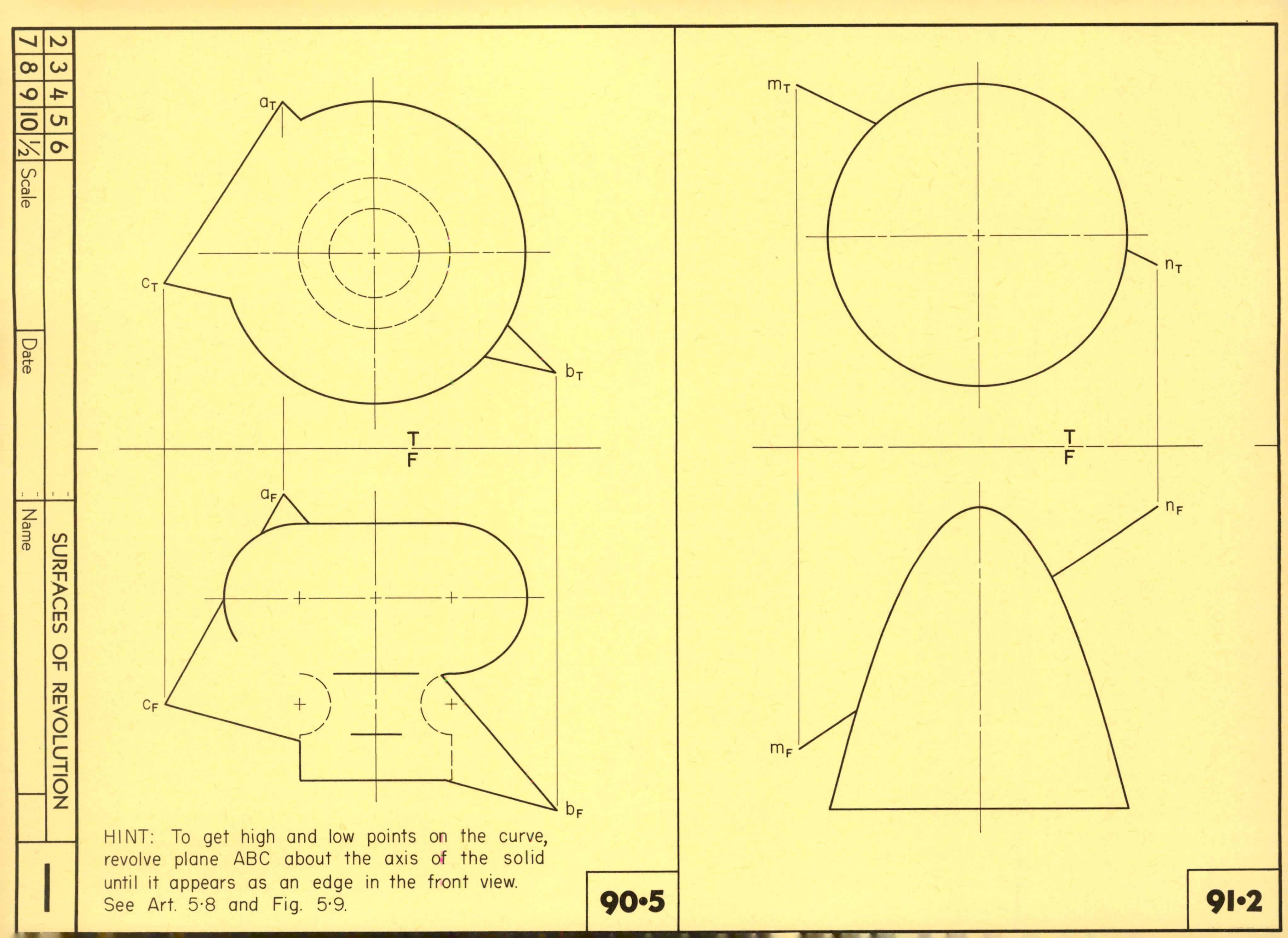

a_T
c_T
b_T
T
F
a_F
c_F
b_F
HINT: To get high and low points on the curve, revolve plane ABC about the axis of the solid until it appears as an edge in the front view. See Art. 5·8 and Fig. 5·9.
90·5
m_T
n_T
n_F
m_F
91·2
2 3 4 5 6
7 8 9 10 ½
Scale
Date
Name
SURFACES OF REVOLUTION

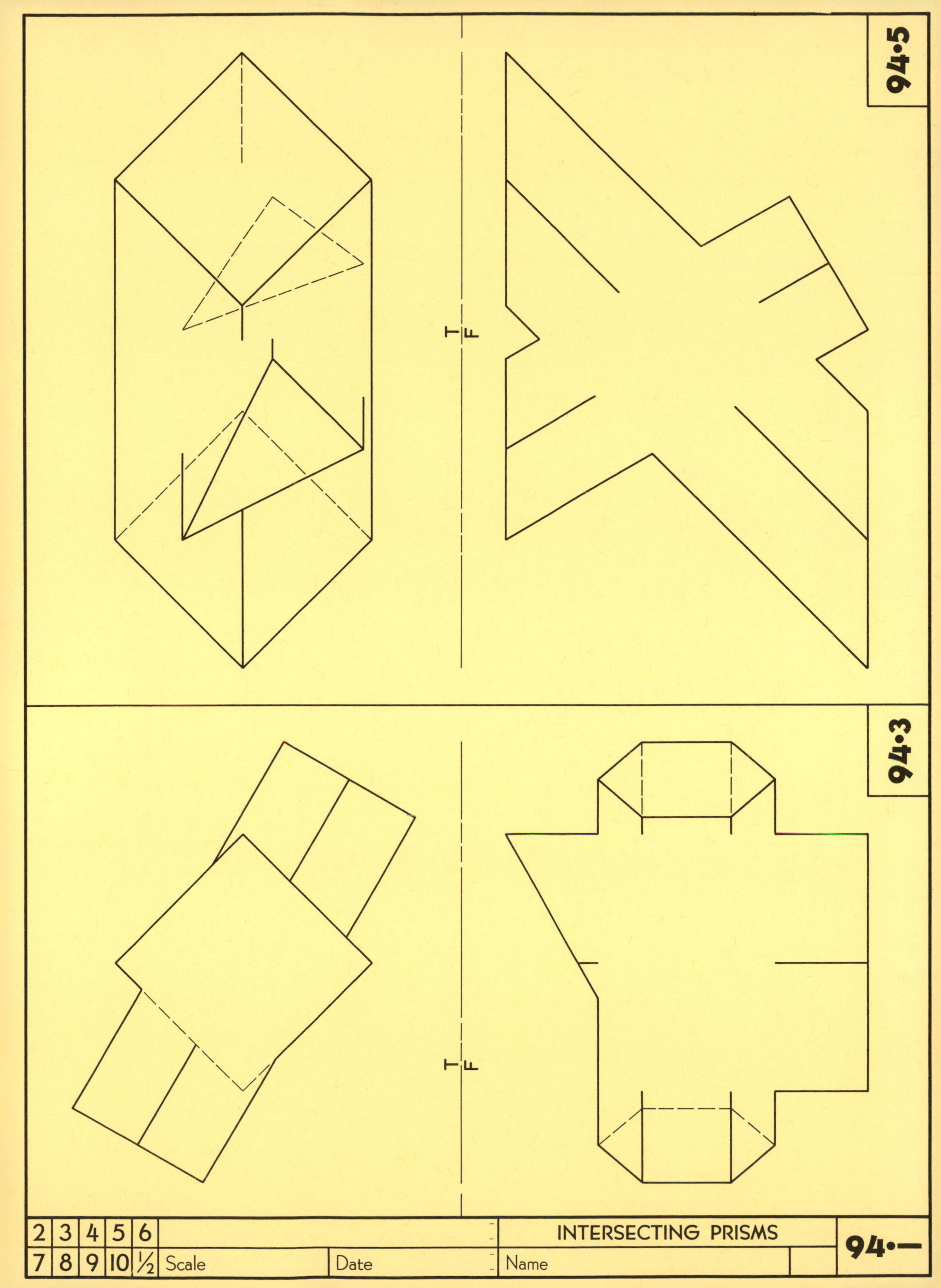
94•5
T
F
94•3
T
F
2 3 4 5 6
7 8 9 10 ½
Scale
Date
INTERSECTING PRISMS
Name
94•—

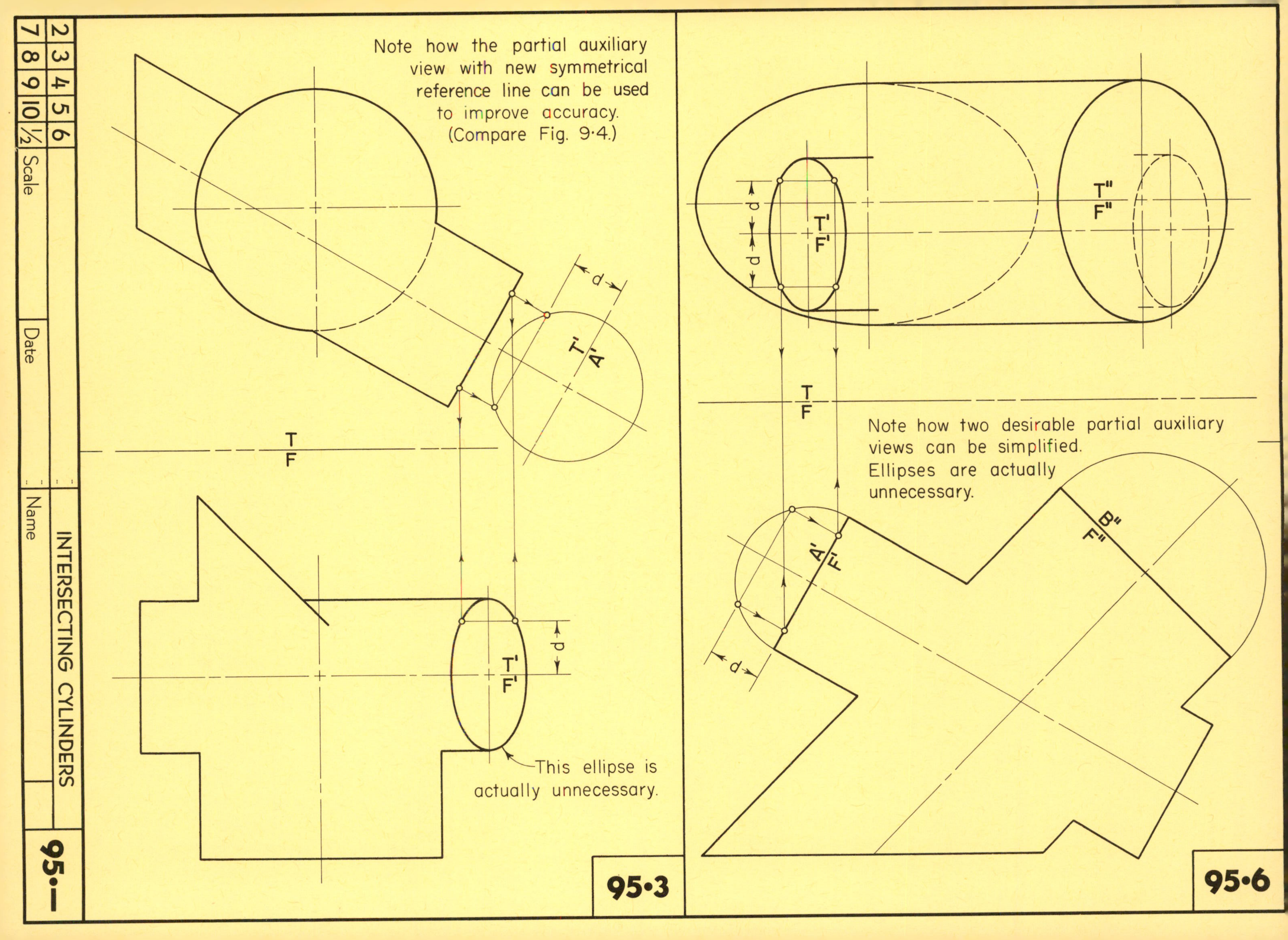
Note how the partial auxiliary view with new symmetrical reference line can be used to improve accuracy. (Compare Fig. 9·4.)
d
T'
A'
T
F
d
T'
F'
This ellipse is actually unnecessary.
95·3
d
d
T'
F'
T"
F"
T
F
Note how two desirable partial auxiliary views can be simplified. Ellipses are actually unnecessary.
B"
F"
A'
F'
d
95·6
2 3 4 5 6
7 8 9 10 ½
Scale
Date
Name
INTERSECTING CYLINDERS
95·—

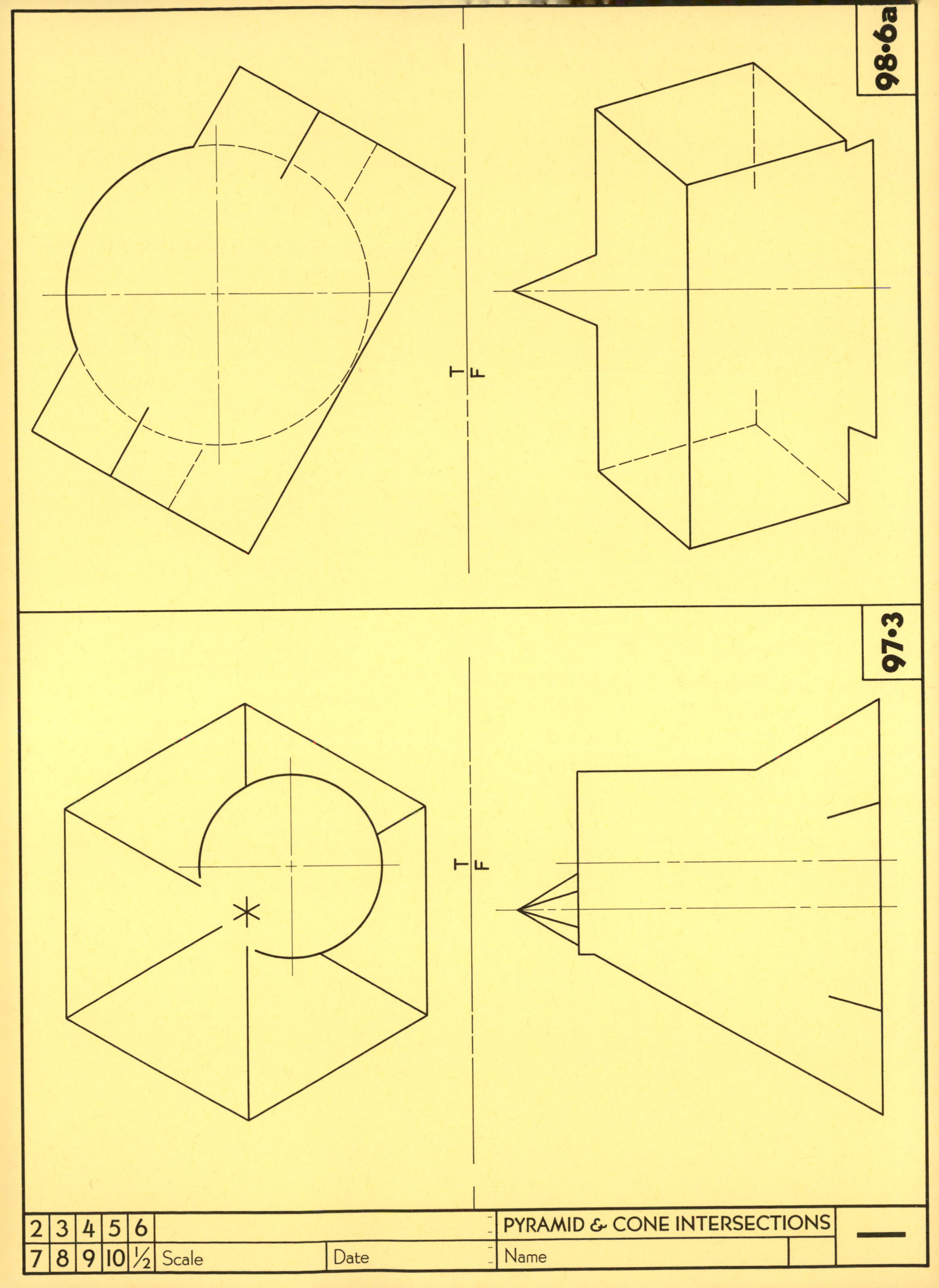
98·6a
T
F
97·3
T
F
2 3 4 5 6
7 8 9 10 ½
Scale
Date
PYRAMID & CONE INTERSECTIONS
Name

T
F
Use the method shown in Fig. 9·11.
2 3 4 5 6
7 8 9 10 ½
Scale
Date
INTERSECTING CYLINDERS
Name
99·6

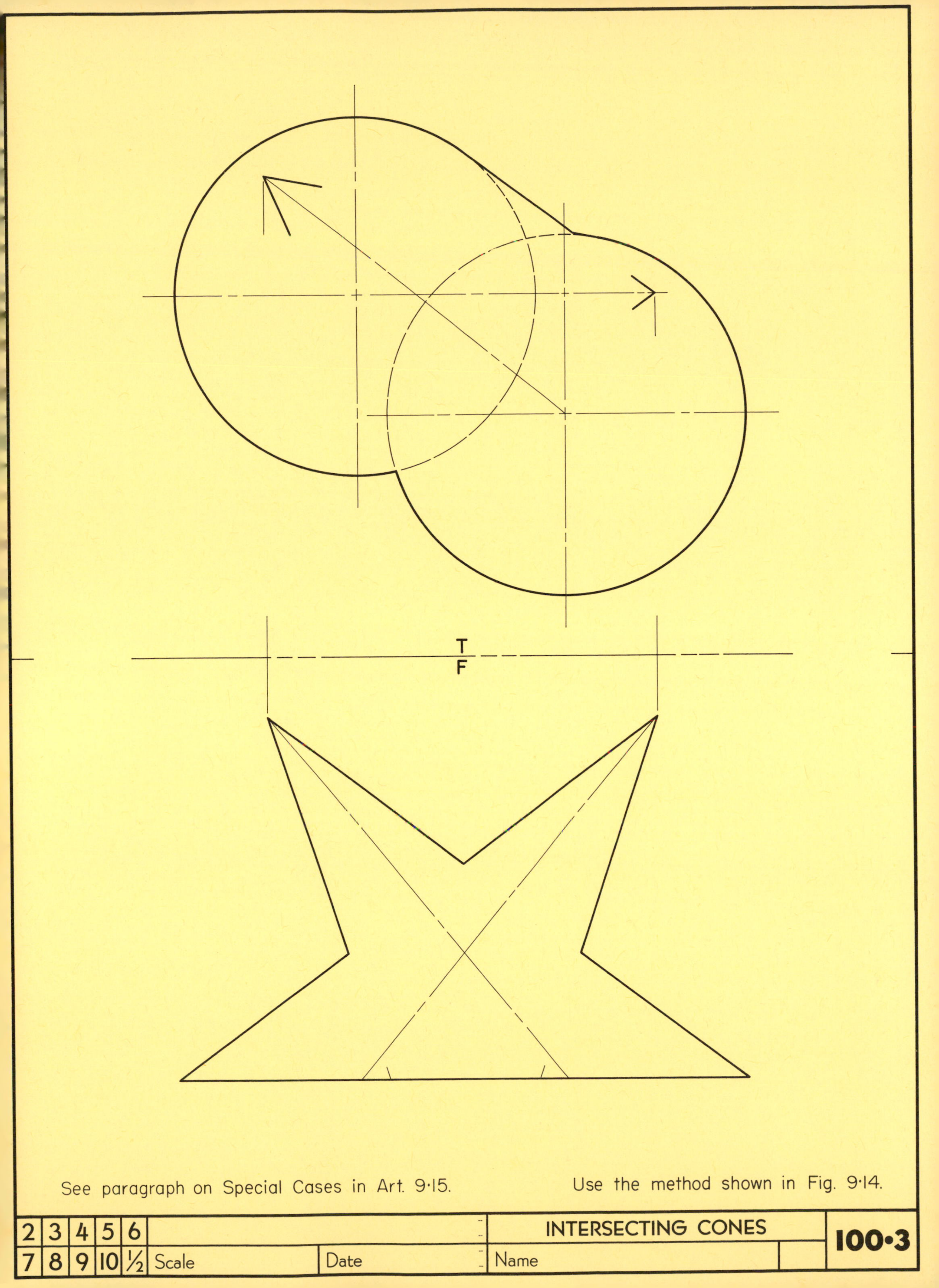
T
F
See paragraph on Special Cases in Art. 9·15.
Use the method shown in Fig. 9·14.
2 3 4 5 6
7 8 9 10 ½
Scale
Date
INTERSECTING CONES
Name
100•3

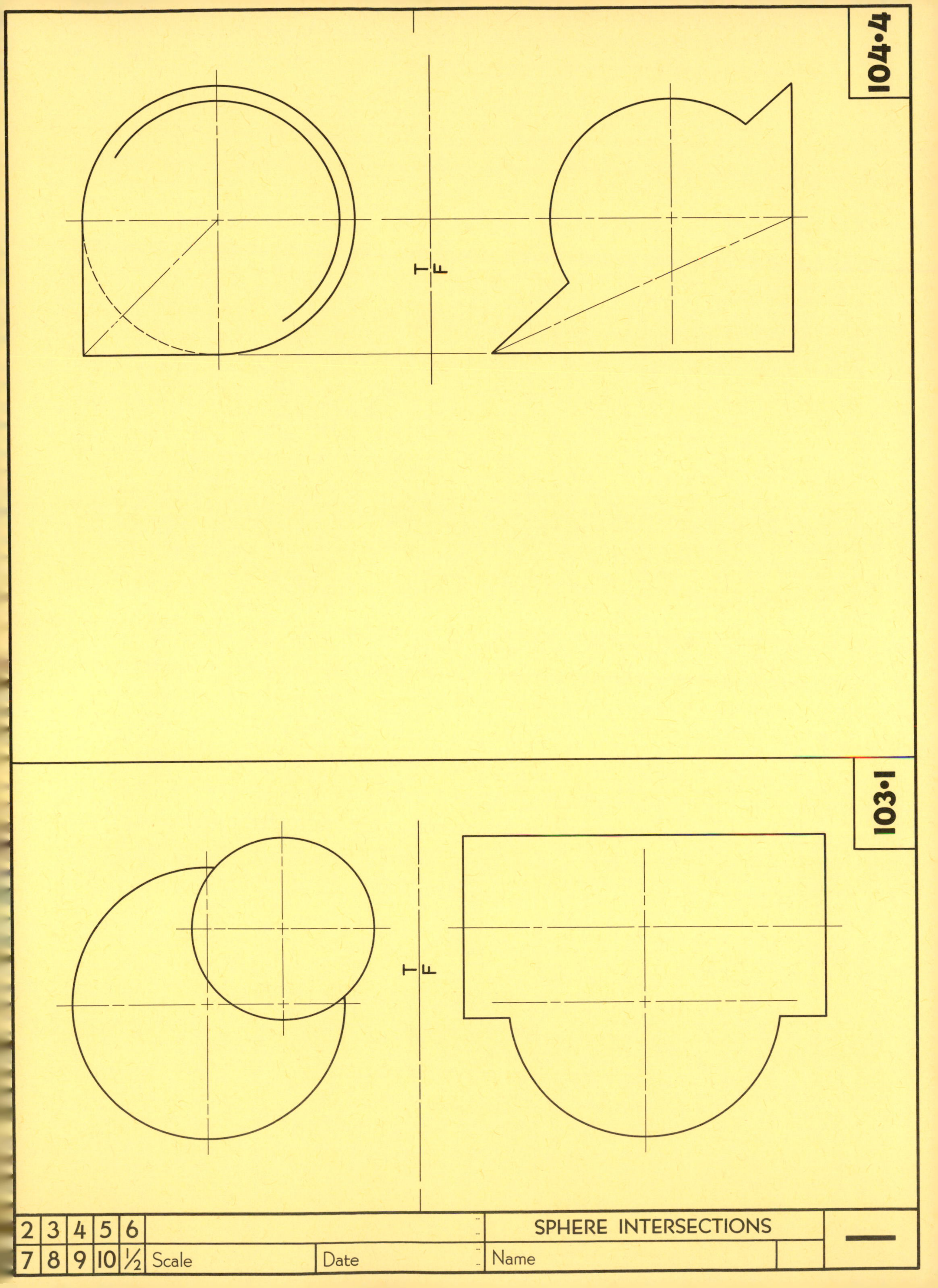

104·4
T
F
103·1
T
F
2 3 4 5 6
7 8 9 10 ½
Scale
Date
SPHERE INTERSECTIONS
Name

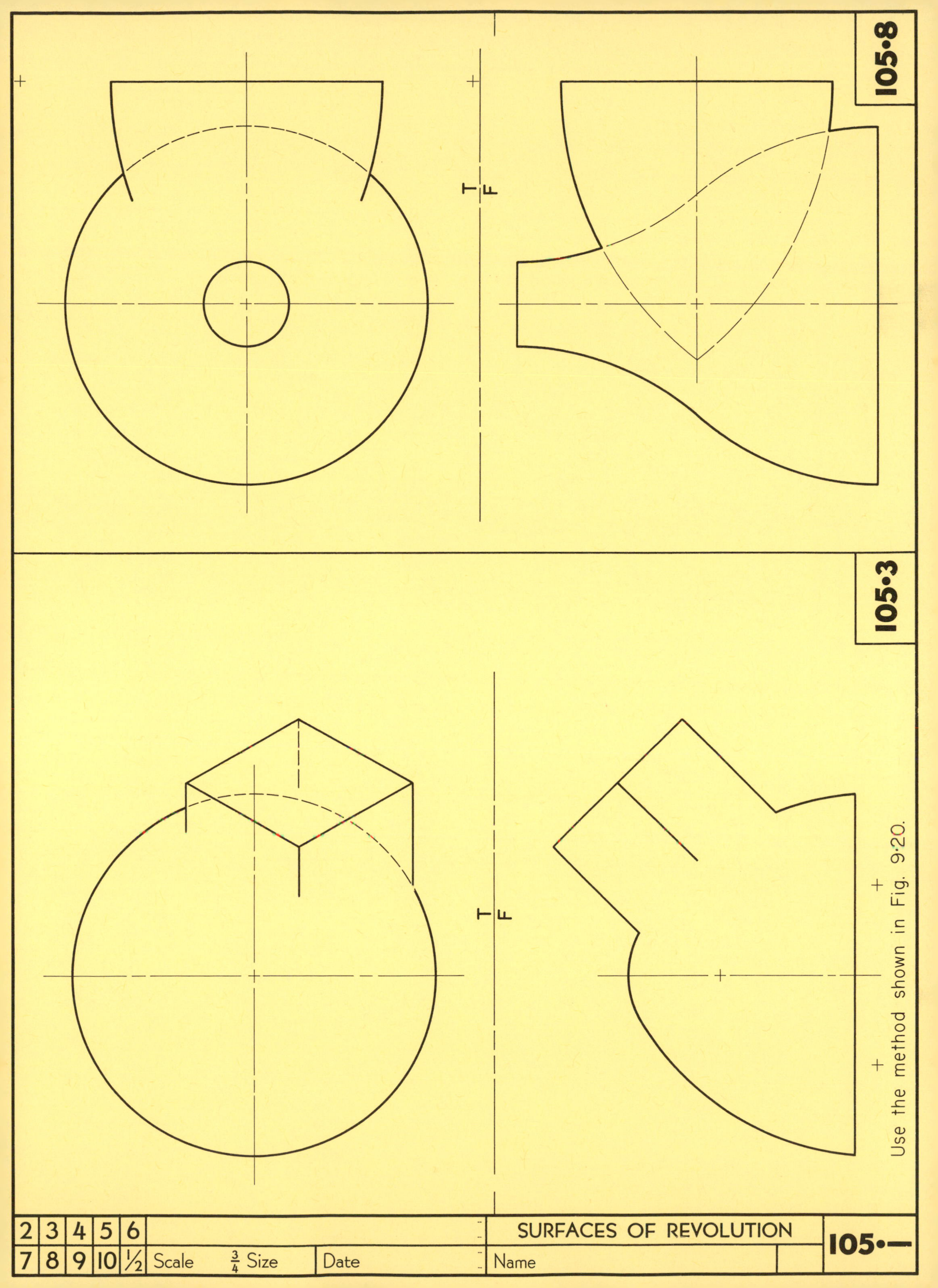
105•8
T
F
105•3
T
F
Use the method shown in Fig. 9·20.
2 3 4 5 6
7 8 9 10 ½ Scale ¾ Size Date
SURFACES OF REVOLUTION
Name
105•—

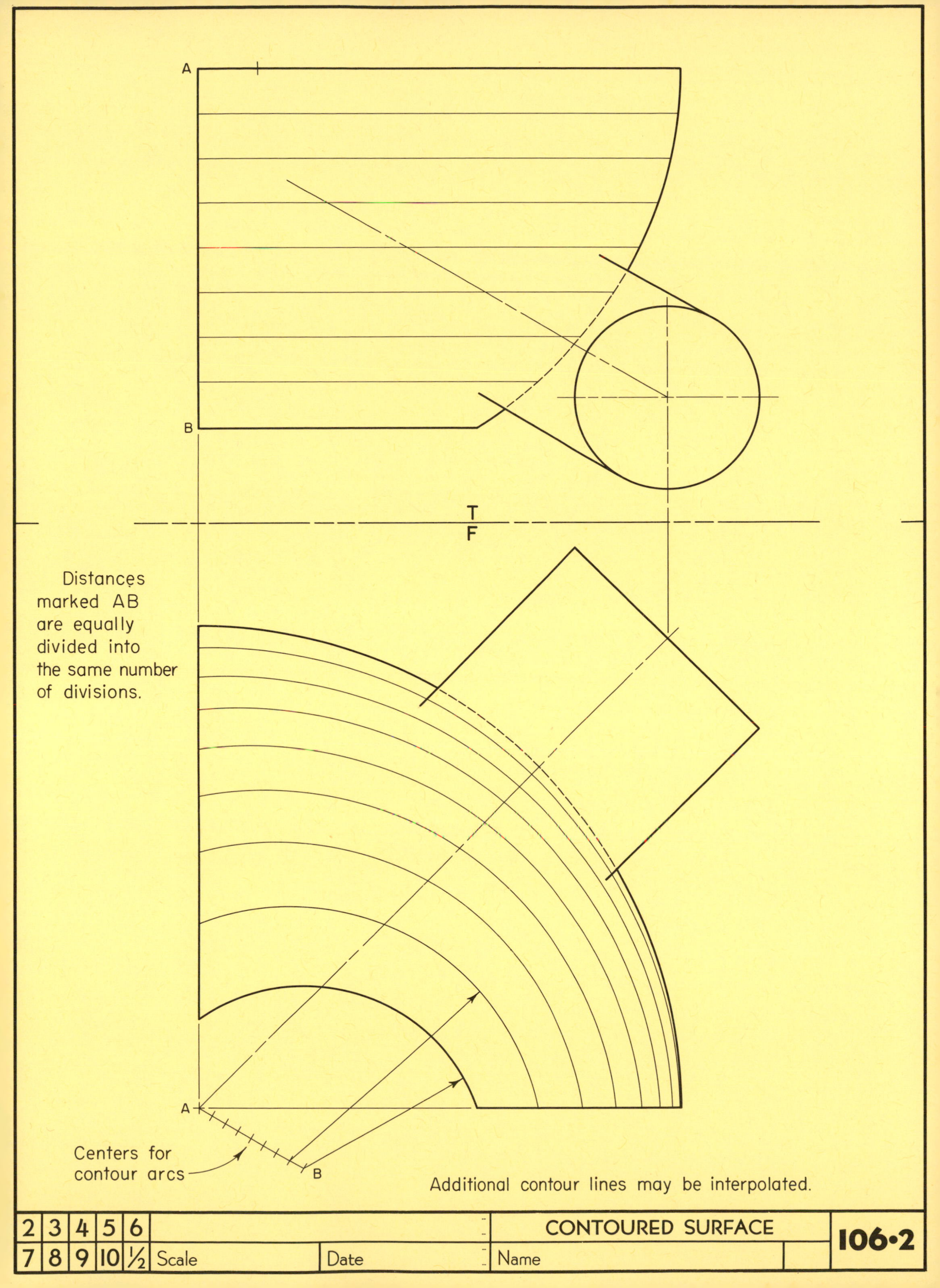
A
B
T
F
Distances marked AB are equally divided into the same number of divisions.
A
Centers for contour arcs
B
Additional contour lines may be interpolated.
2 3 4 5 6
7 8 9 10 ½ Scale
Date
CONTOURED SURFACE
Name
106•2

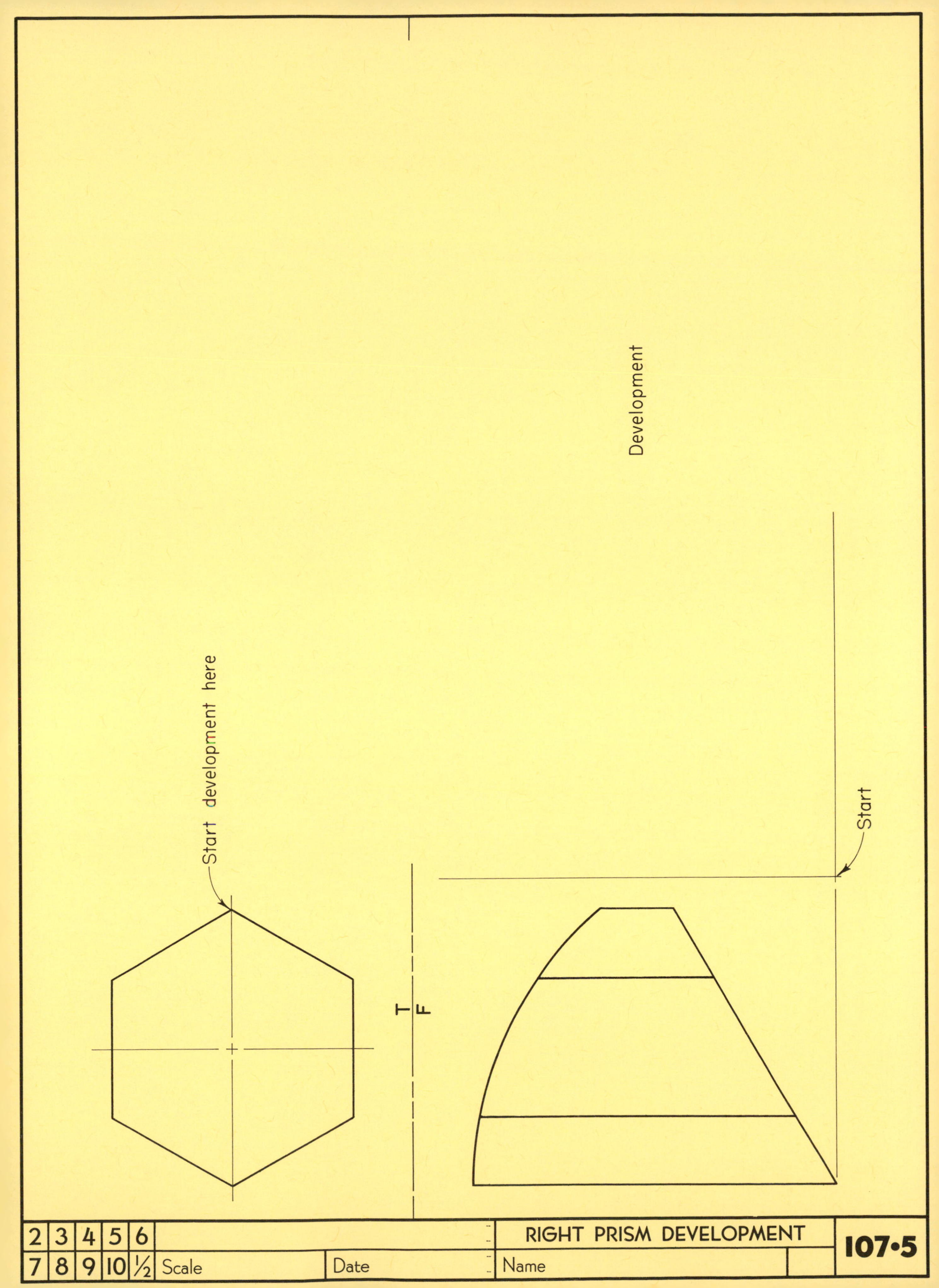
Development
Start development here
Start
T
F
2 3 4 5 6
7 8 9 10 ½
Scale
Date
Name
RIGHT PRISM DEVELOPMENT
107•5

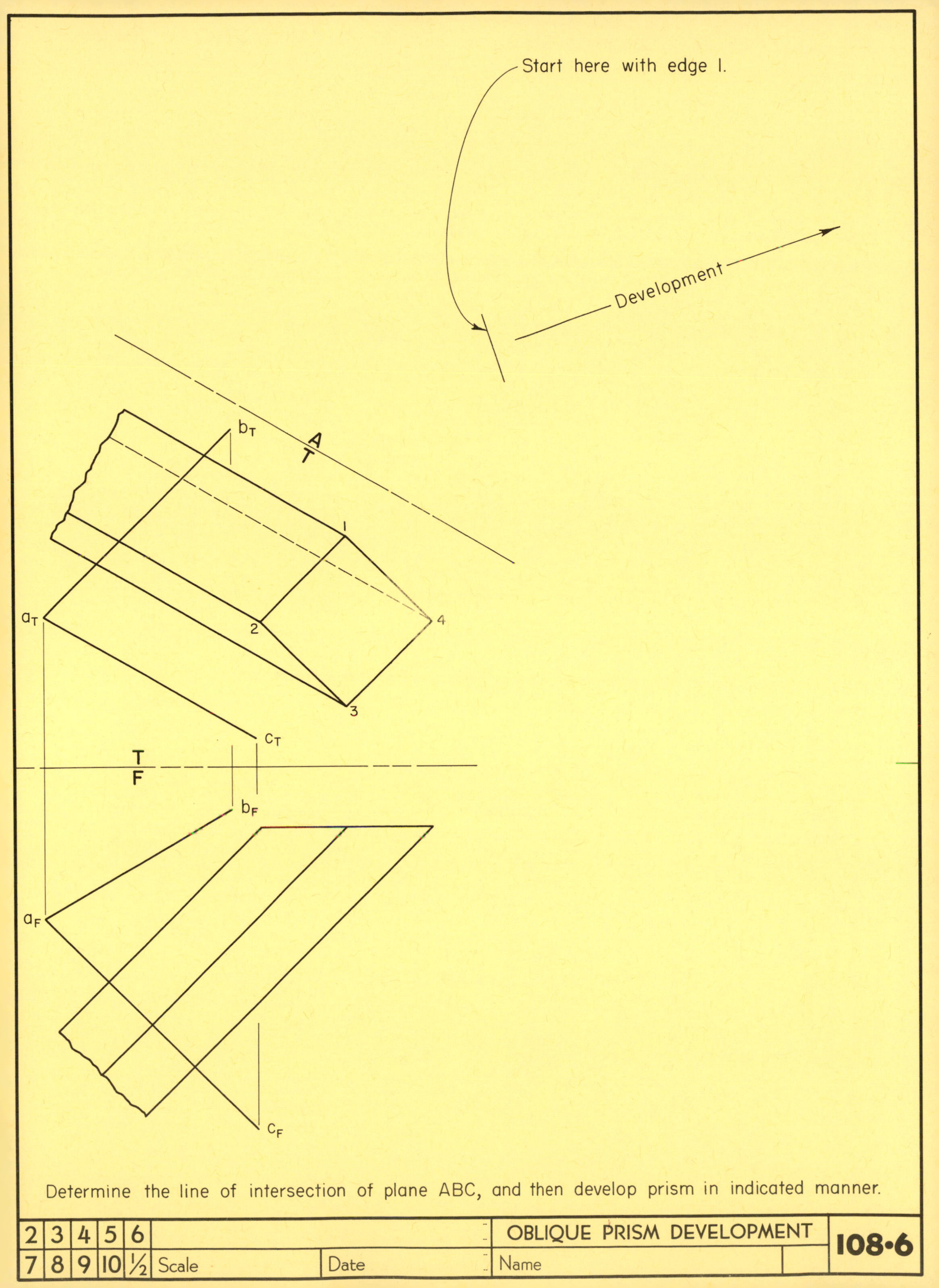

Determine the line of intersection of plane ABC, and then develop prism in indicated manner.

2	3	4	5	6		OBLIQUE PRISM DEVELOPMENT	108·6
7	8	9	10	½	Scale / Date	Name	

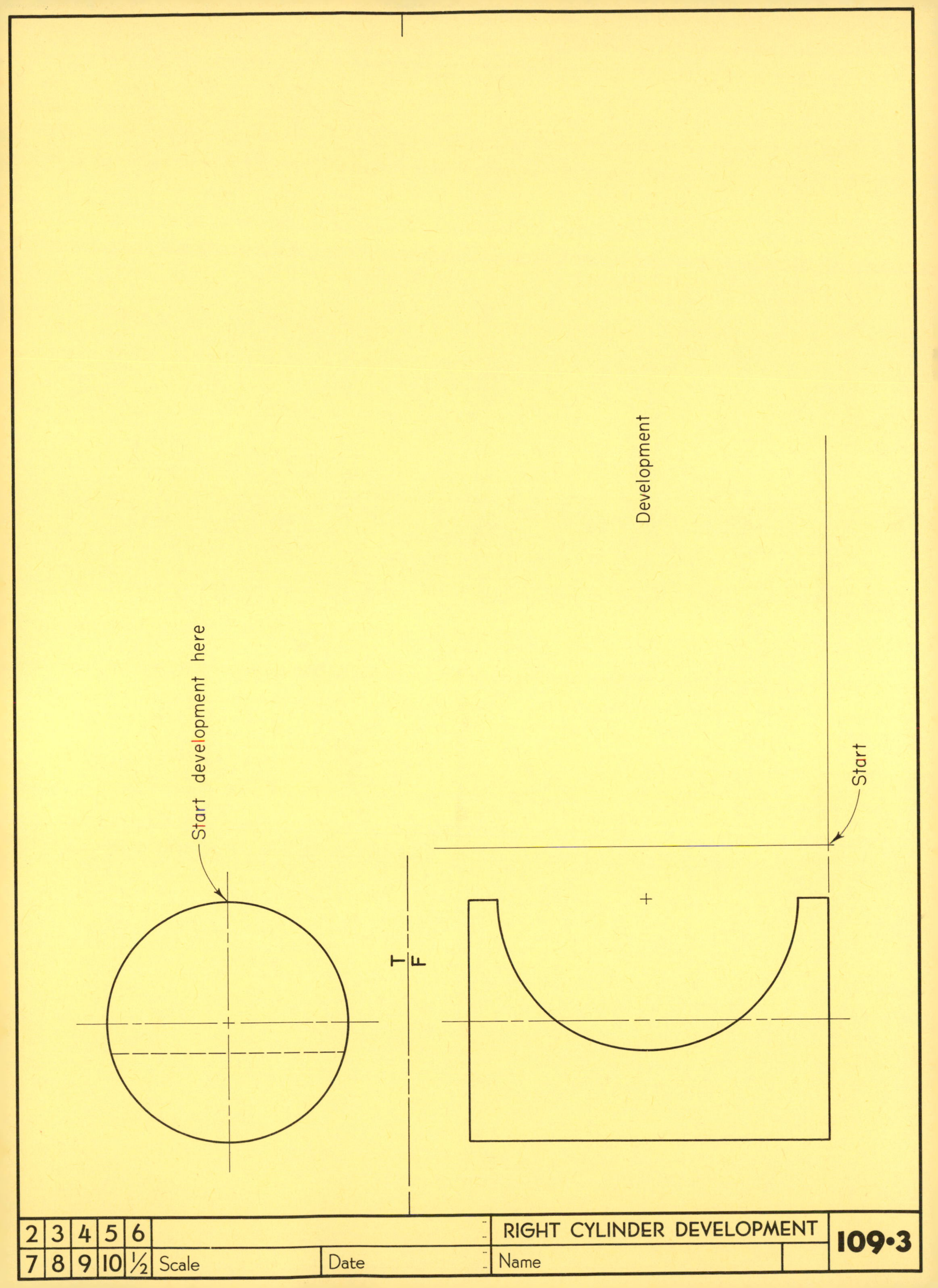
Start development here
Development
Start
T
F
2 3 4 5 6
7 8 9 10 ½
Scale
Date
RIGHT CYLINDER DEVELOPMENT
Name
109·3

Draw necessary views on this sheet, but draw the development on a second sheet aligned with the true-length view. Start with element O.

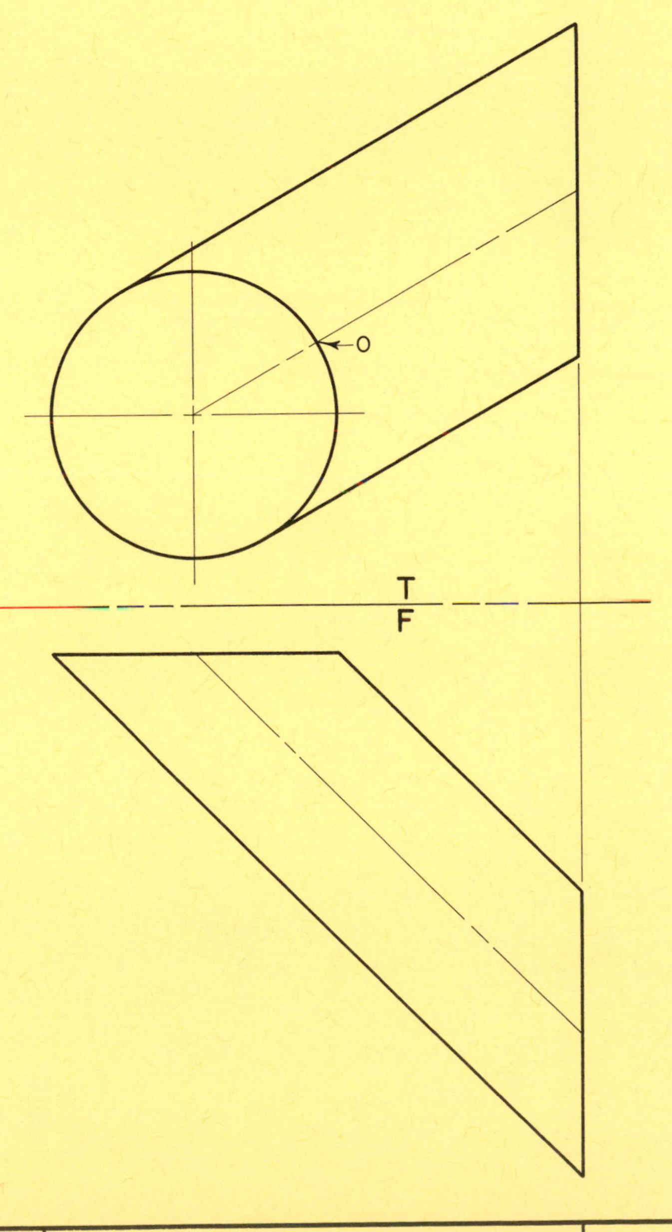

2	3	4	5	6		OBLIQUE CYLINDER DEVELOPMENT	110•2
7	8	9	10	½	Scale / Date	Name	

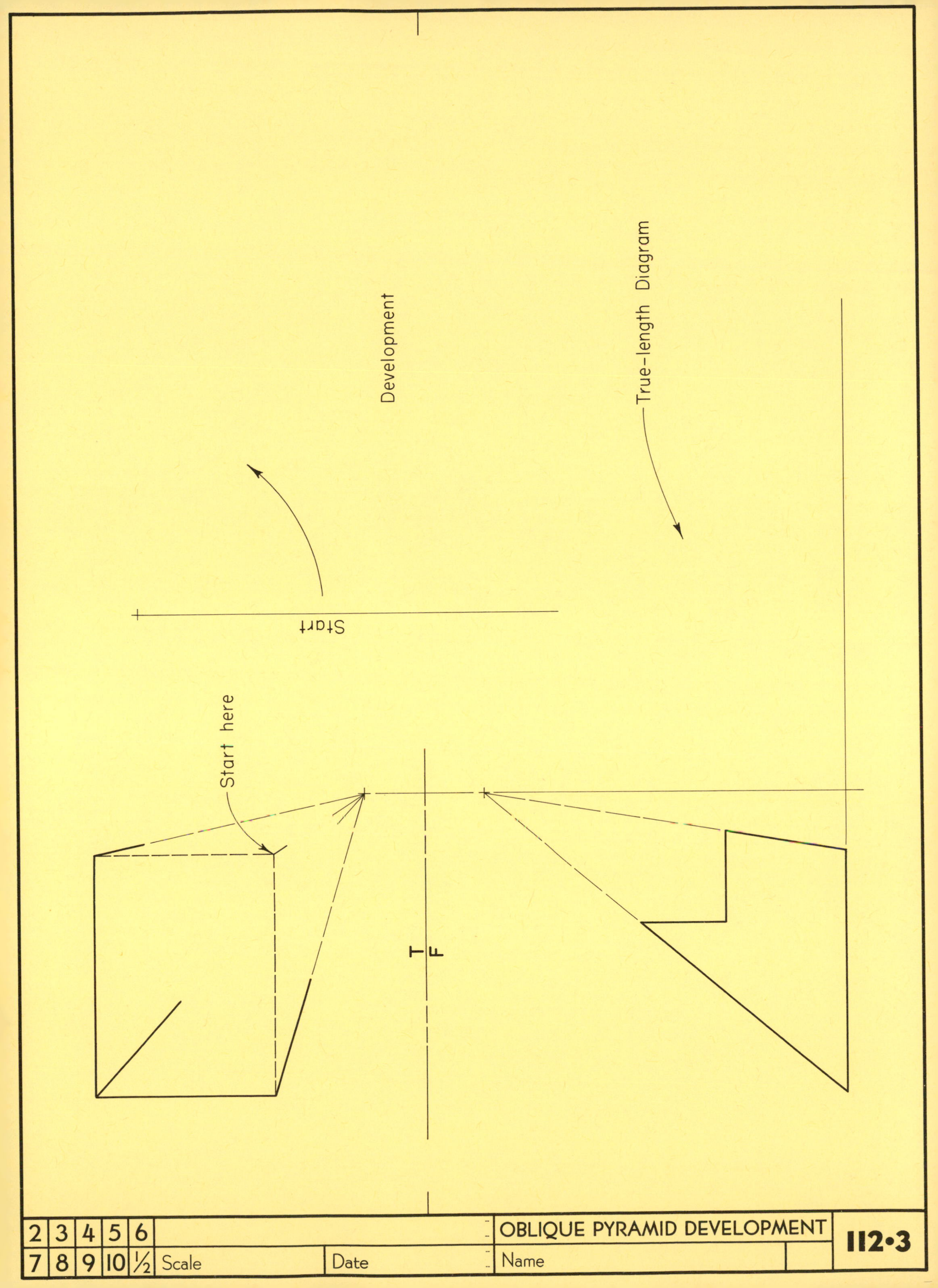
Development
True-length Diagram
Start
Start here
T
F
2 3 4 5 6
7 8 9 10 ½
Scale
Date
OBLIQUE PYRAMID DEVELOPMENT
Name
112•3

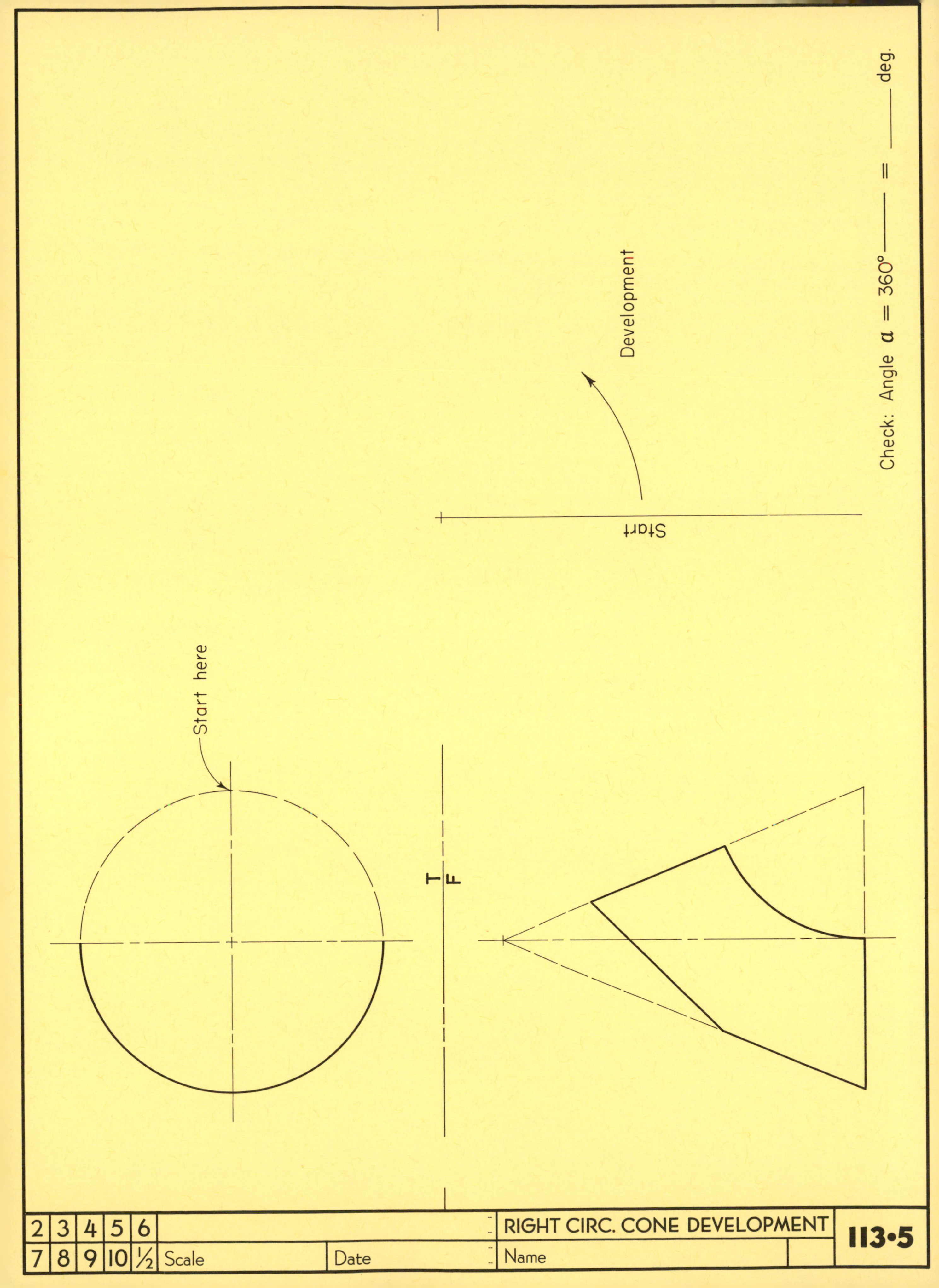

Start here
T
F
Start
Development
Check: Angle a = 360° ____ = ____ deg.
2 3 4 5 6
7 8 9 10 ½
Scale
Date
RIGHT CIRC. CONE DEVELOPMENT
Name
113•5

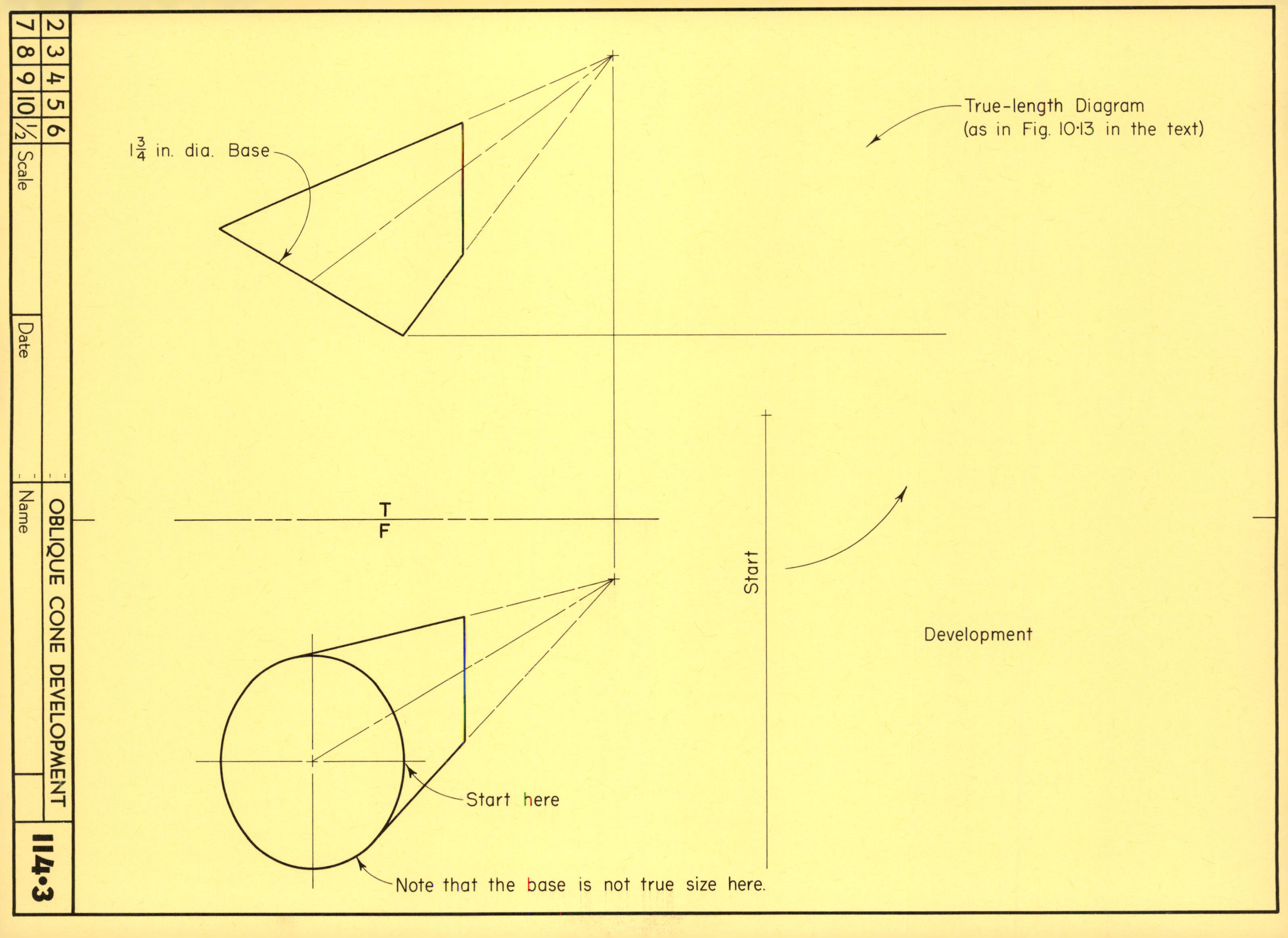
True-length Diagram
(as in Fig. 10·13 in the text)
$1\frac{3}{4}$ in. dia. Base
T
F
Start
Development
Start here
Note that the base is not true size here.
2 3 4 5 6
7 8 9 10 ½
Scale
Date
Name
OBLIQUE CONE DEVELOPMENT
114•3

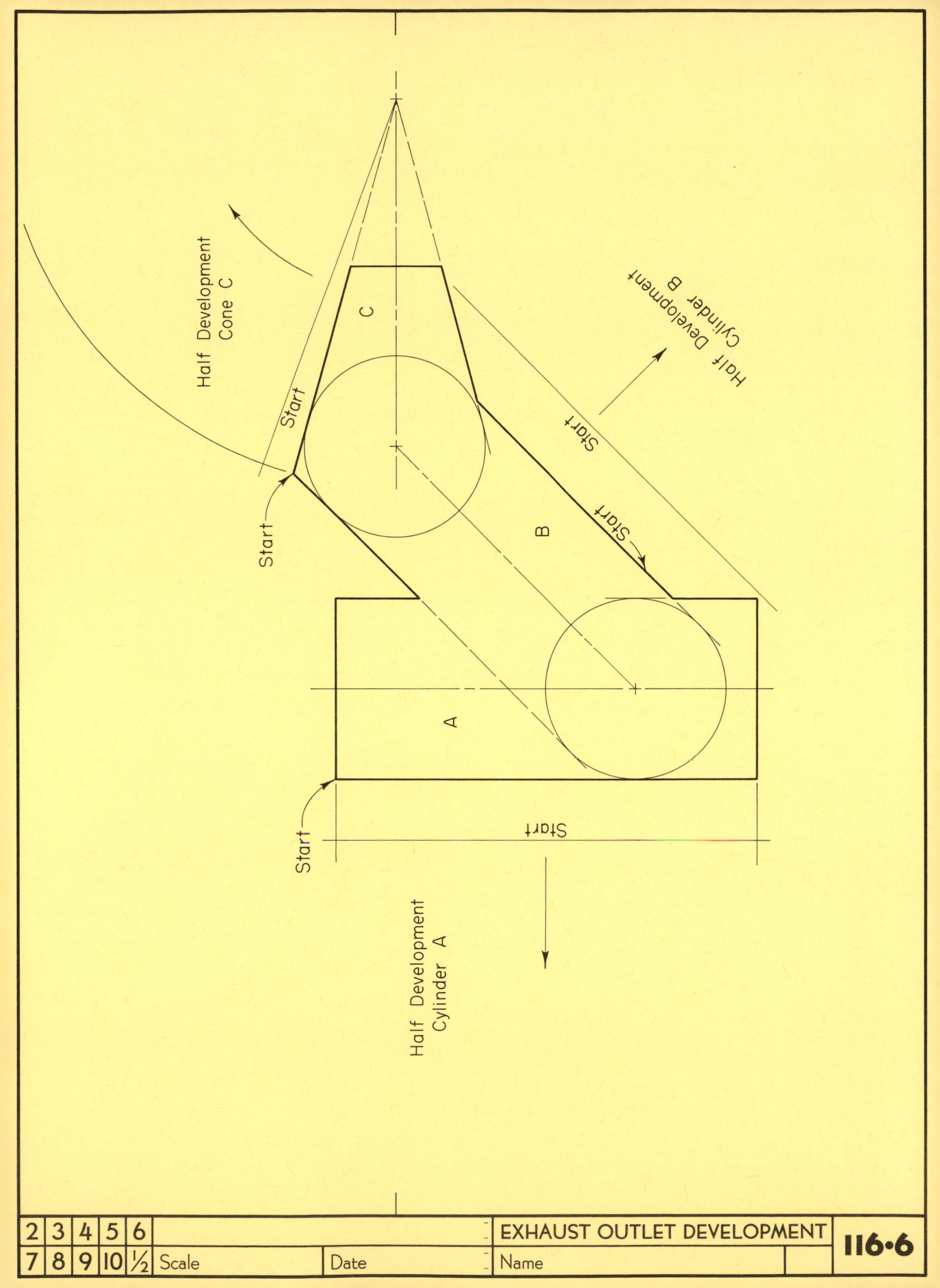
Half Development
Cone C
C
Start
Start
Half Development
Cylinder B
Start
Start
B
A
Start
Start
Half Development
Cylinder A
2 3 4 5 6
7 8 9 10 ½
Scale
Date
EXHAUST OUTLET DEVELOPMENT
Name
116·6

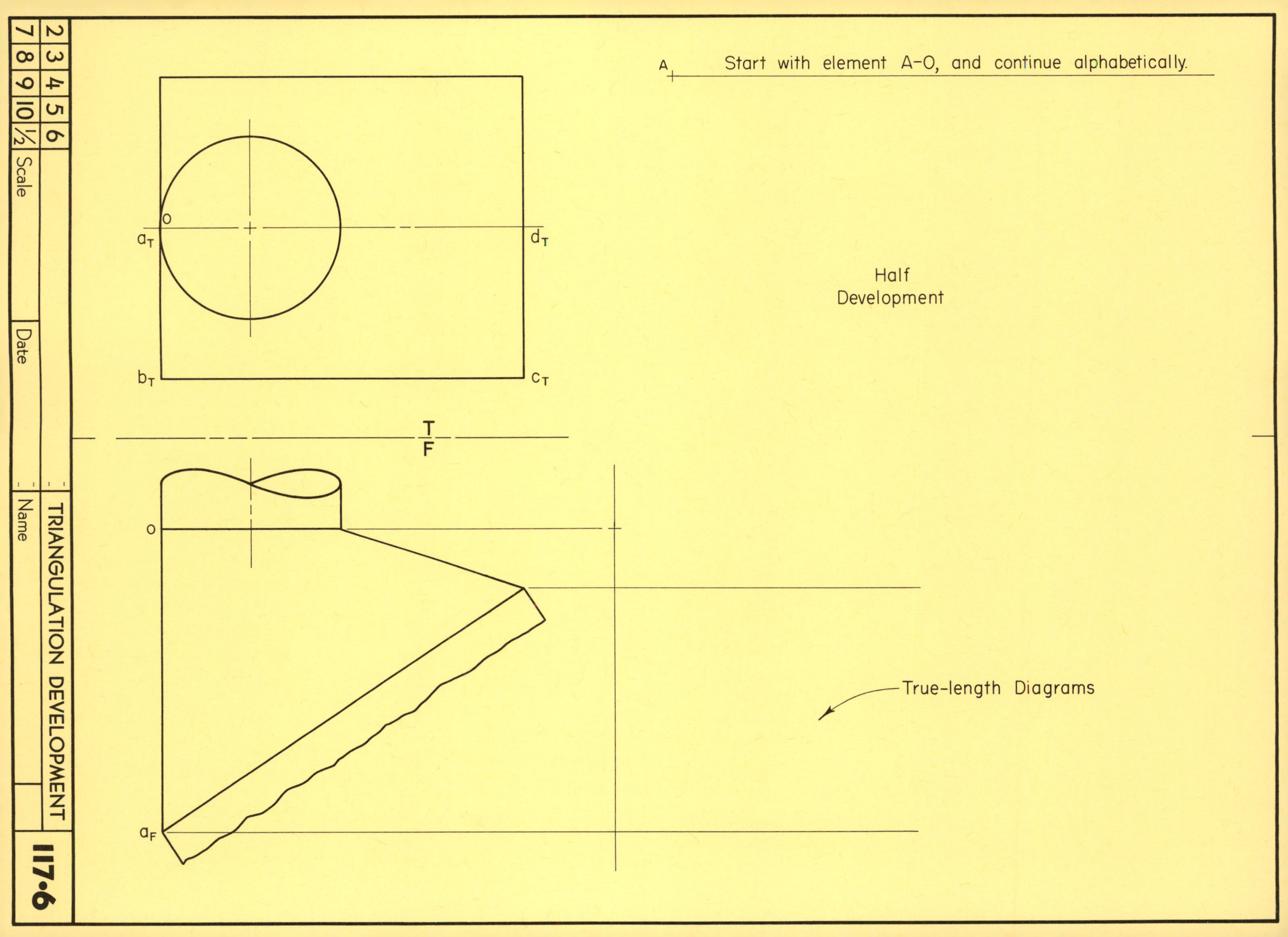
Start with element A-O, and continue alphabetically.
A
Half
Development
True-length Diagrams
T
F
O
a_T
d_T
b_T
c_T
O
a_F
2 3 4 5 6
7 8 9 10 ½ Scale
Date
TRIANGULATION DEVELOPMENT
Name
117·6

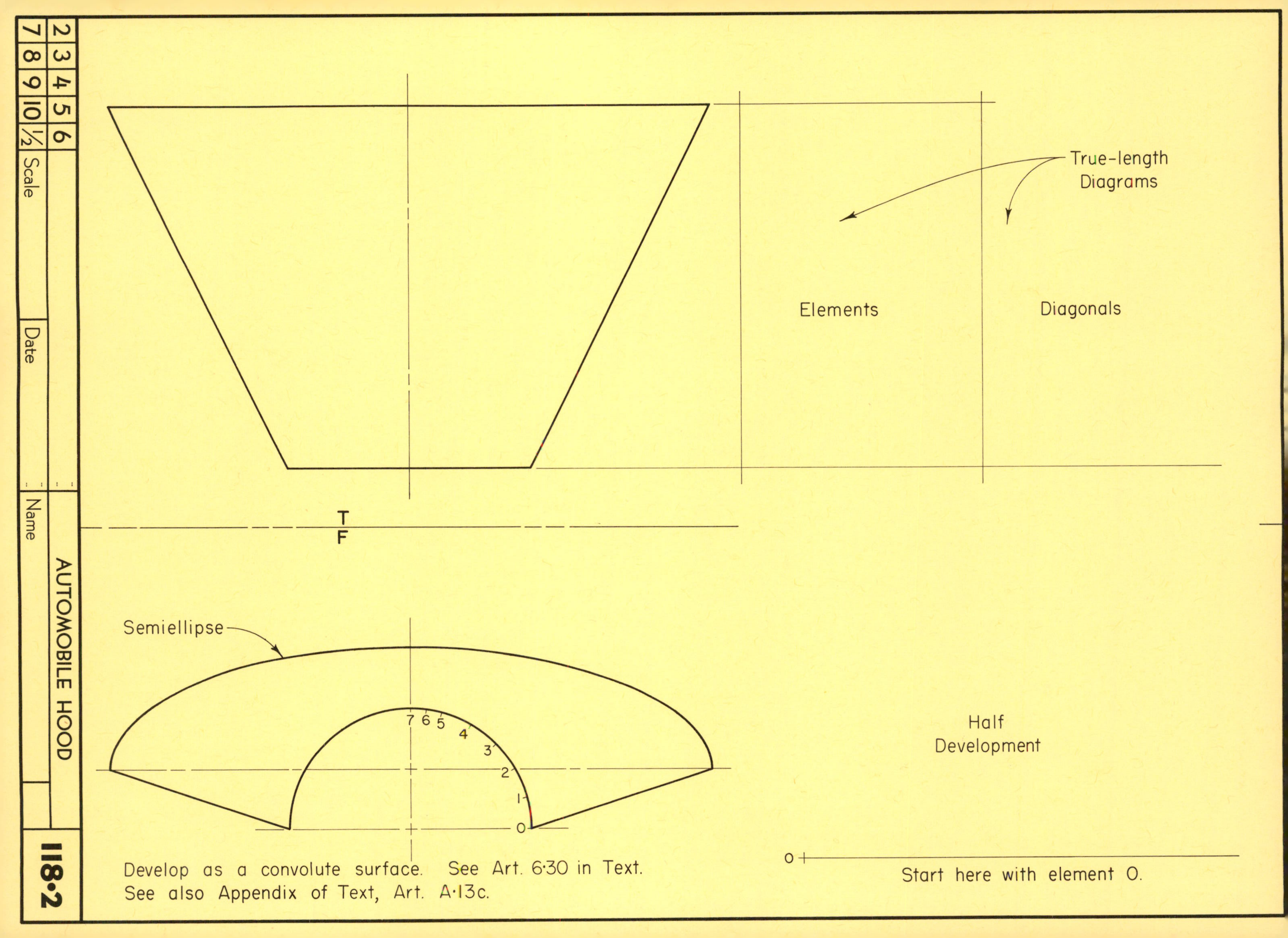

True-length
Diagrams
Elements
Diagonals
T
F
Semiellipse
7
6
5
4
3
2
1
0
Half
Development
0
Start here with element O.
Develop as a convolute surface. See Art. 6·30 in Text.
See also Appendix of Text, Art. A·13c.
2 3 4 5 6
7 8 9 10 ½
Scale
Date
Name
AUTOMOBILE HOOD
118·2

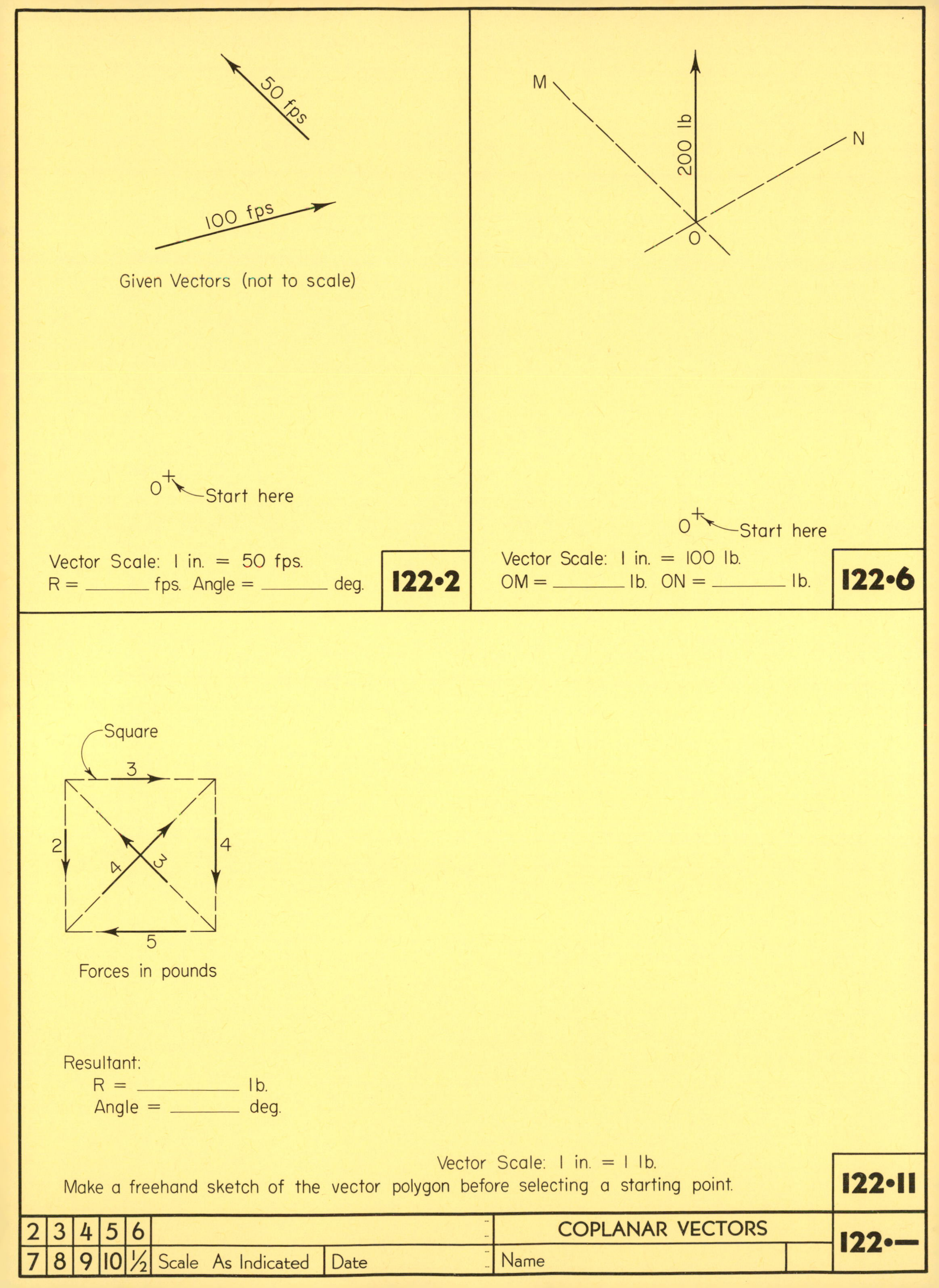
50 fps
100 fps
Given Vectors (not to scale)
O
Start here
Vector Scale: 1 in. = 50 fps.
R = ______ fps. Angle = ______ deg.
122•2
M
200 lb
N
O
O
Start here
Vector Scale: 1 in. = 100 lb.
OM = ______ lb. ON = ______ lb.
122•6
Square
3
2
4
4
3
5
Forces in pounds
Resultant:
R = ______ lb.
Angle = ______ deg.
Vector Scale: 1 in. = 1 lb.
Make a freehand sketch of the vector polygon before selecting a starting point.
122•11
2 3 4 5 6
7 8 9 10 ½
Scale As Indicated
Date
COPLANAR VECTORS
Name
122•—

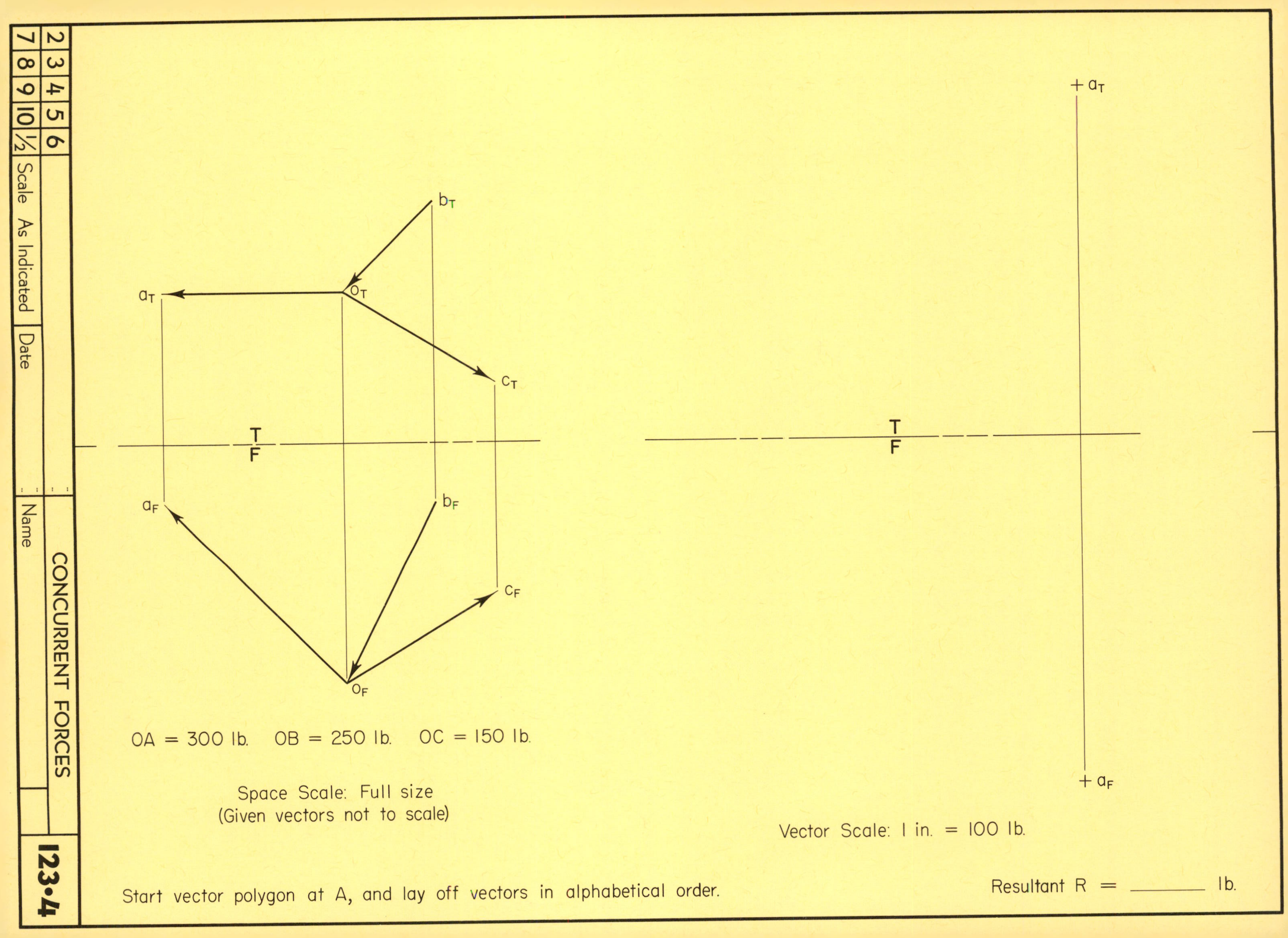

b_T
a_T
o_T
c_T
T
F
a_F
b_F
c_F
o_F
OA = 300 lb. OB = 250 lb. OC = 150 lb.
Space Scale: Full size
(Given vectors not to scale)
a_T
T
F
a_F
Vector Scale: 1 in. = 100 lb.
Start vector polygon at A, and lay off vectors in alphabetical order.
Resultant R = ______ lb.
2 3 4 5 6
7 8 9 10 ½
Scale As Indicated
Date
Name
CONCURRENT FORCES
123•4

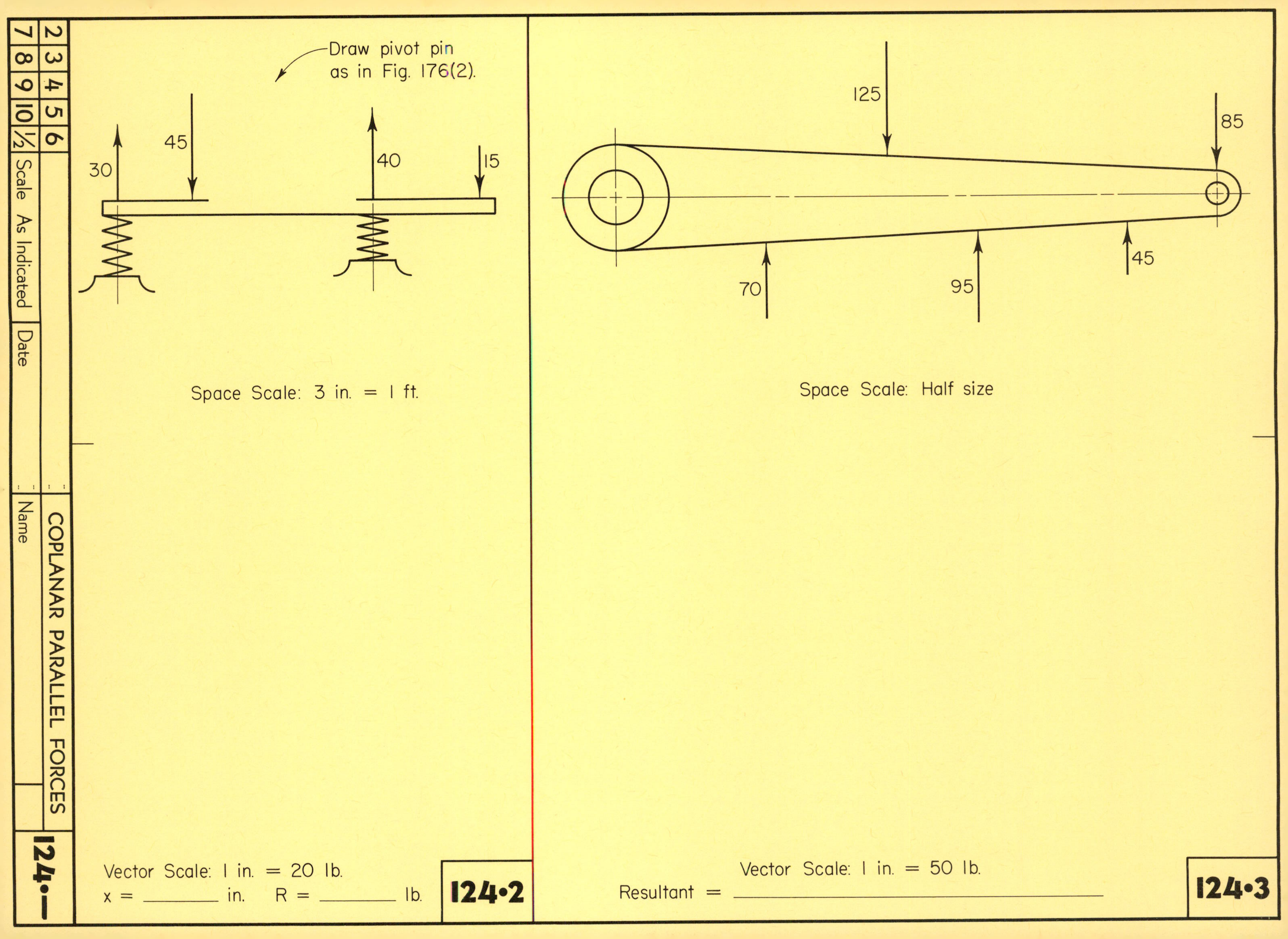
Draw pivot pin
as in Fig. 176(2).
30
45
40
15
Space Scale: 3 in. = 1 ft.
Vector Scale: 1 in. = 20 lb.
x = ____ in. R = ____ lb.
124•2
125
85
70
95
45
Space Scale: Half size
Vector Scale: 1 in. = 50 lb.
Resultant =
124•3
2 3 4 5 6
7 8 9 10 ½
Scale As Indicated
Date
COPLANAR PARALLEL FORCES
Name
124•—

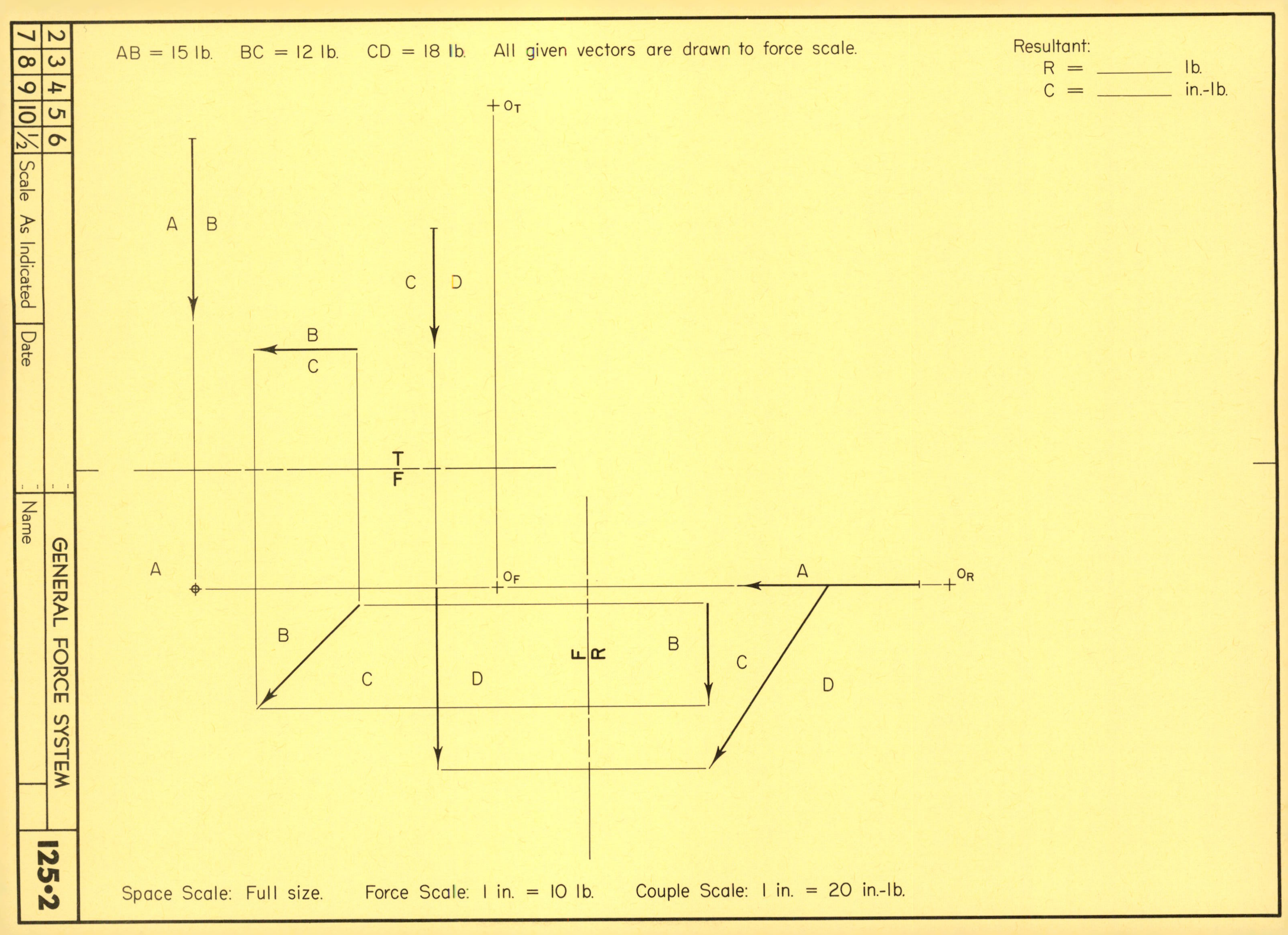
AB = 15 lb.
BC = 12 lb.
CD = 18 lb.
All given vectors are drawn to force scale.
Resultant:
R = ______ lb.
C = ______ in.-lb.
A
B
C
D
T
F
R
O_T
O_F
O_R
Space Scale: Full size.
Force Scale: 1 in. = 10 lb.
Couple Scale: 1 in. = 20 in.-lb.
2 3 4 5 6
7 8 9 10 ½
Scale As Indicated
Date
Name
GENERAL FORCE SYSTEM
125•2

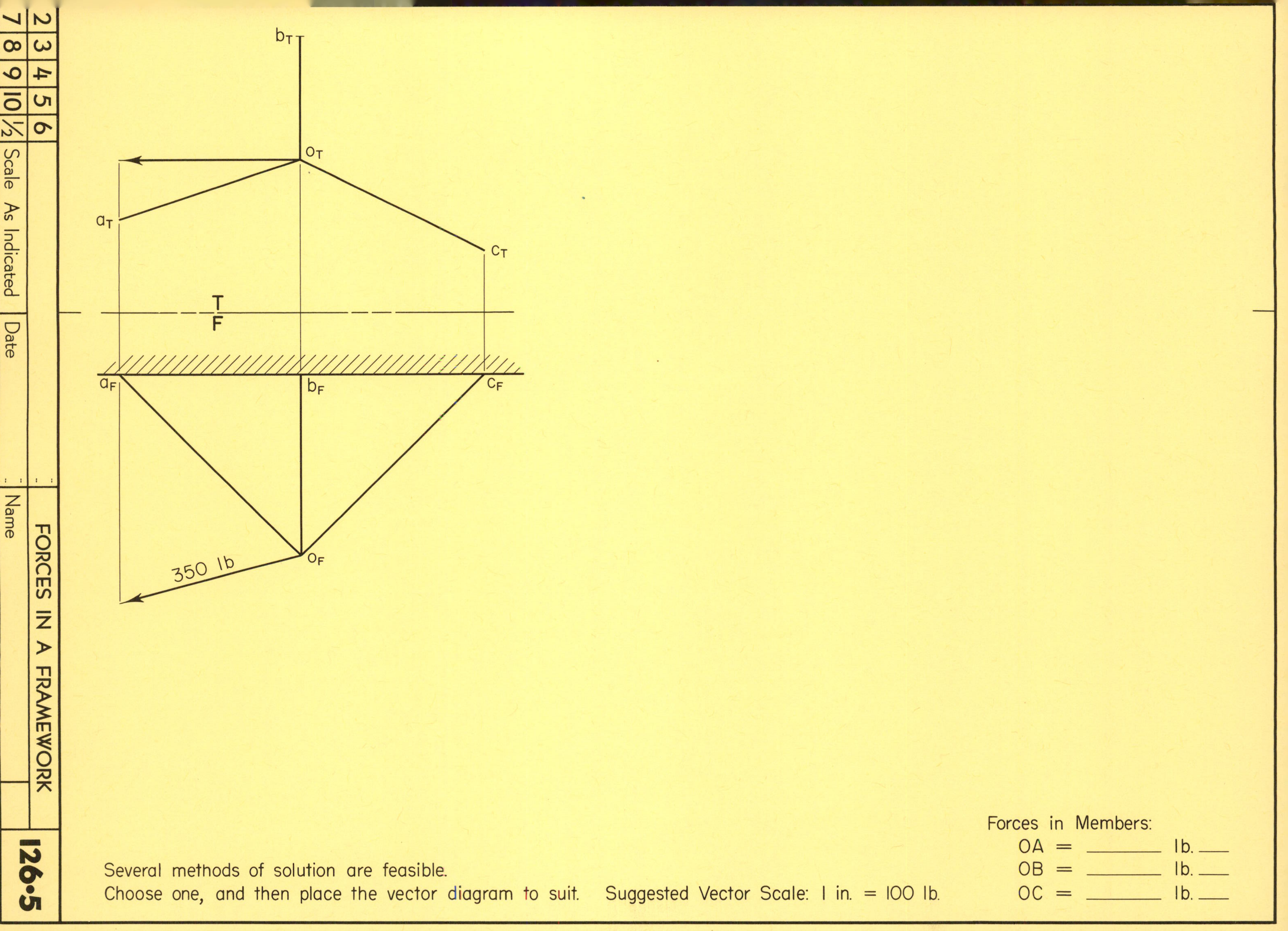

Several methods of solution are feasible.
Choose one, and then place the vector diagram to suit. Suggested Vector Scale: 1 in. = 100 lb.

Forces in Members:

OA = ________ lb. ___
OB = ________ lb. ___
OC = ________ lb. ___

2	3	4	5	6		FORCES IN A FRAMEWORK	126•5
7	8	9	10	½	Scale As Indicated	Date	Name

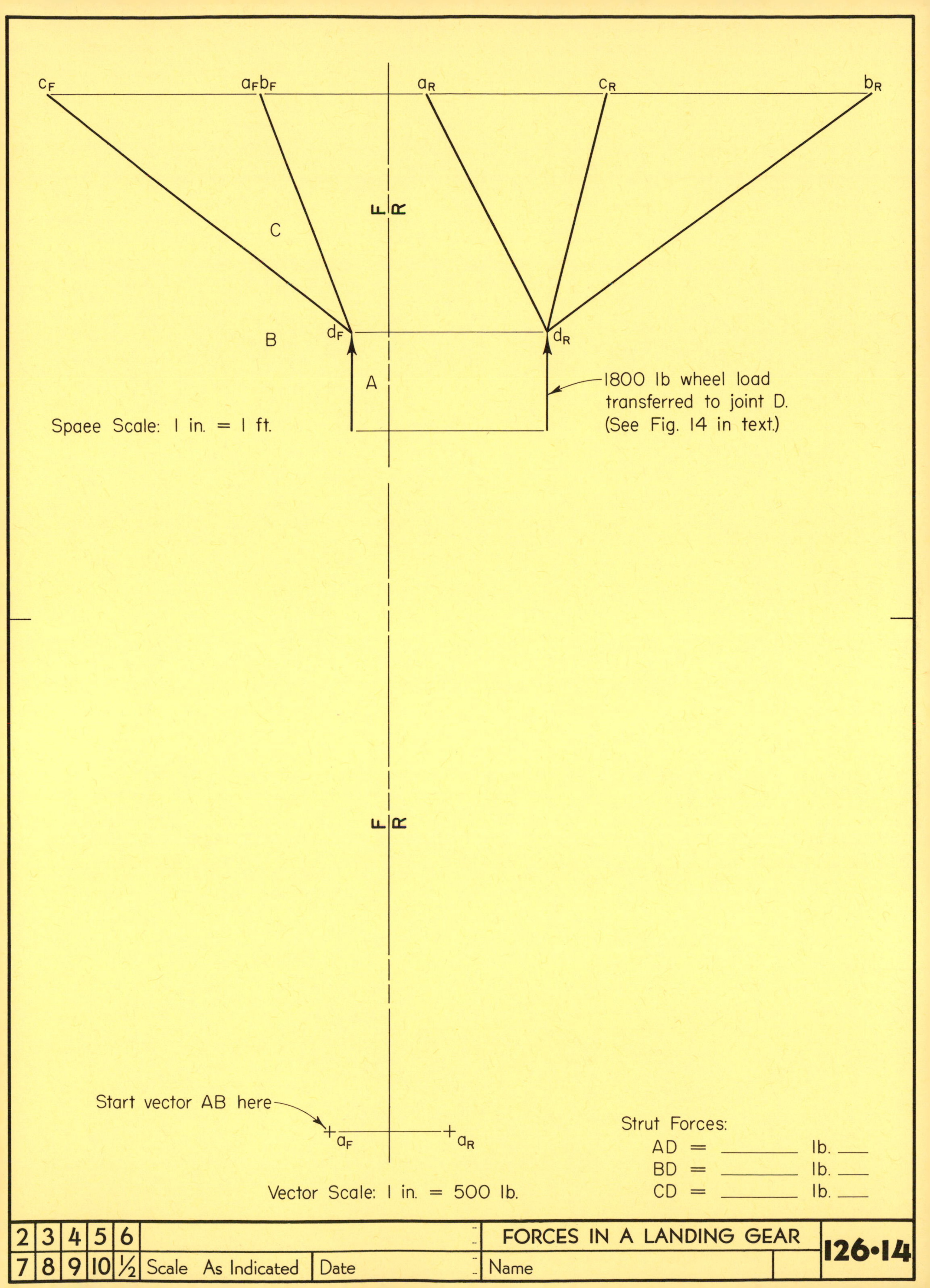

c_F
$a_F b_F$
a_R
c_R
b_R
F
R
C
B
d_F
d_R
A
1800 lb wheel load
transferred to joint D.
(See Fig. 14 in text.)
Spaee Scale: 1 in. = 1 ft.
F
R
Start vector AB here
a_F
a_R
Vector Scale: 1 in. = 500 lb.
Strut Forces:
AD = ________ lb. ___
BD = ________ lb. ___
CD = ________ lb. ___
2 3 4 5 6
7 8 9 10 ½
Scale As Indicated
Date
FORCES IN A LANDING GEAR
Name
126•14

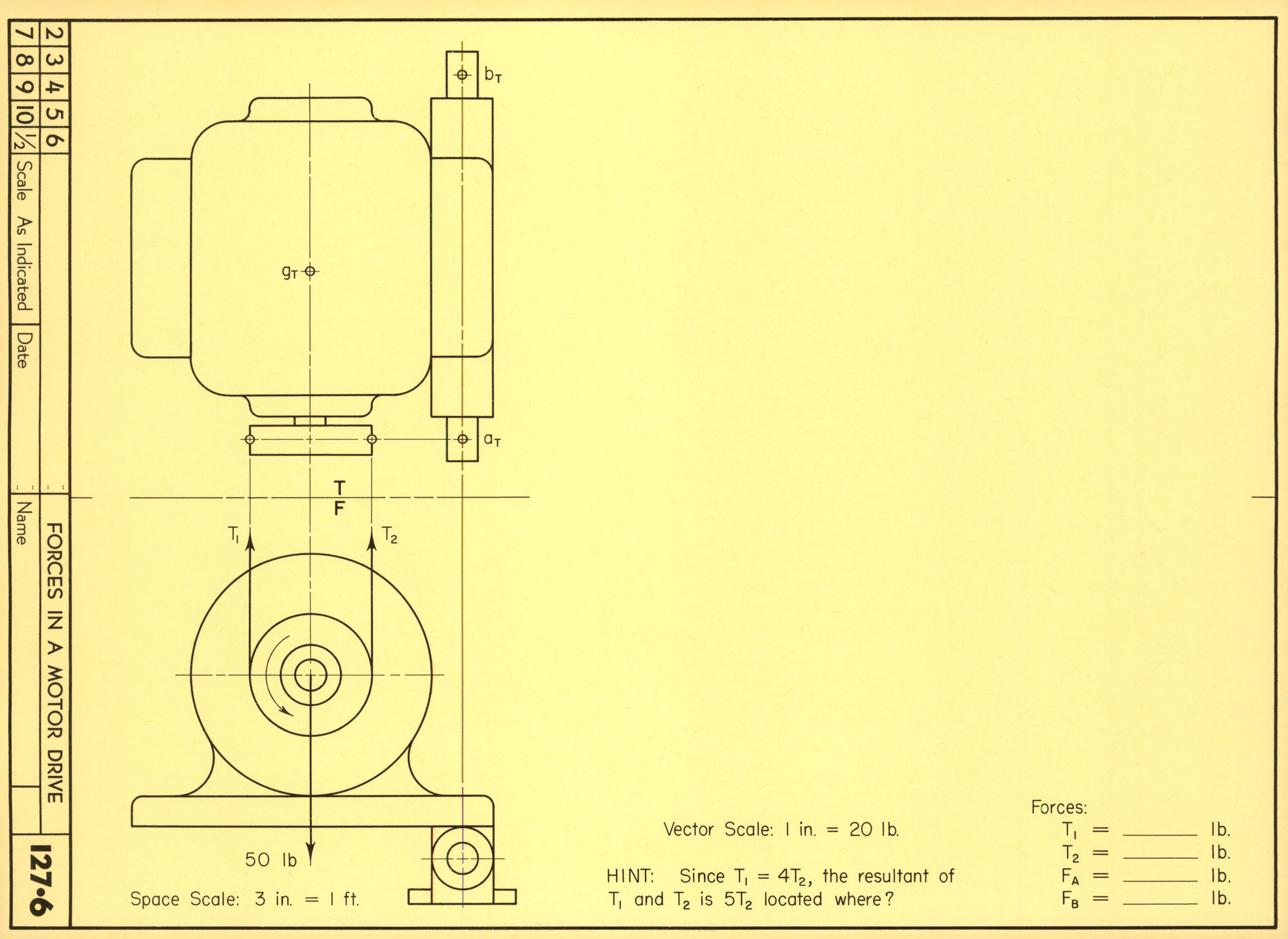

b_T
g_T
a_T
T
F
T_1
T_2
50 lb
Space Scale: 3 in. = 1 ft.
Vector Scale: 1 in. = 20 lb.
HINT: Since $T_1 = 4T_2$, the resultant of T_1 and T_2 is $5T_2$ located where?
Forces:
T_1 = ______ lb.
T_2 = ______ lb.
F_A = ______ lb.
F_B = ______ lb.
2 3 4 5 6
7 8 9 10 ½
Scale As Indicated
Date
FORCES IN A MOTOR DRIVE
Name
127•6

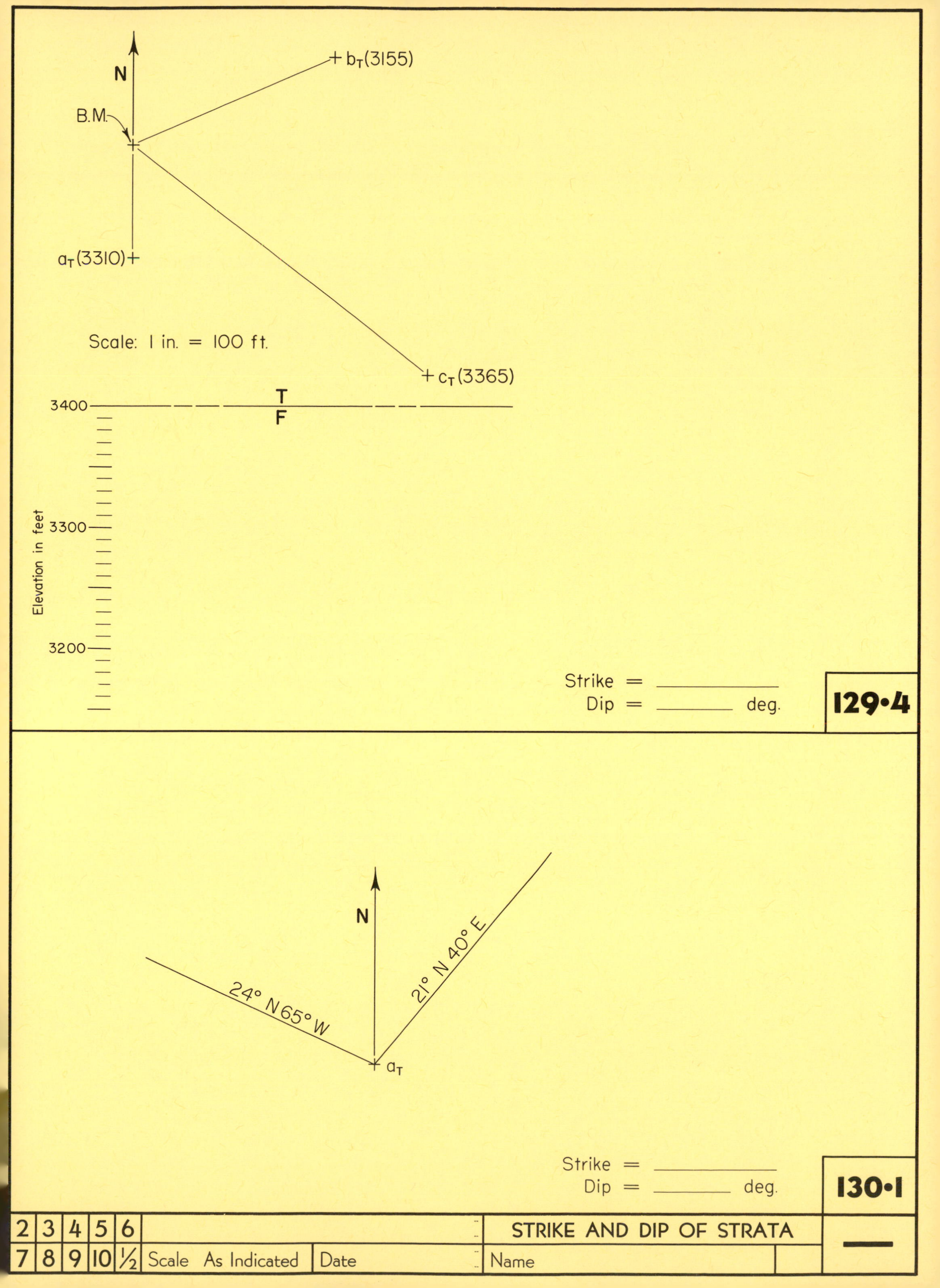

N
B.M.
b_T(3155)
a_T(3310)
c_T(3365)
Scale: 1 in. = 100 ft.
3400
T
F
3300
3200
Elevation in feet
Strike =
Dip = deg.
129•4
N
24° N65° W
21° N 40° E
a_T
Strike =
Dip = deg.
130•1
2 3 4 5 6
7 8 9 10 ½
Scale As Indicated
Date
STRIKE AND DIP OF STRATA
Name

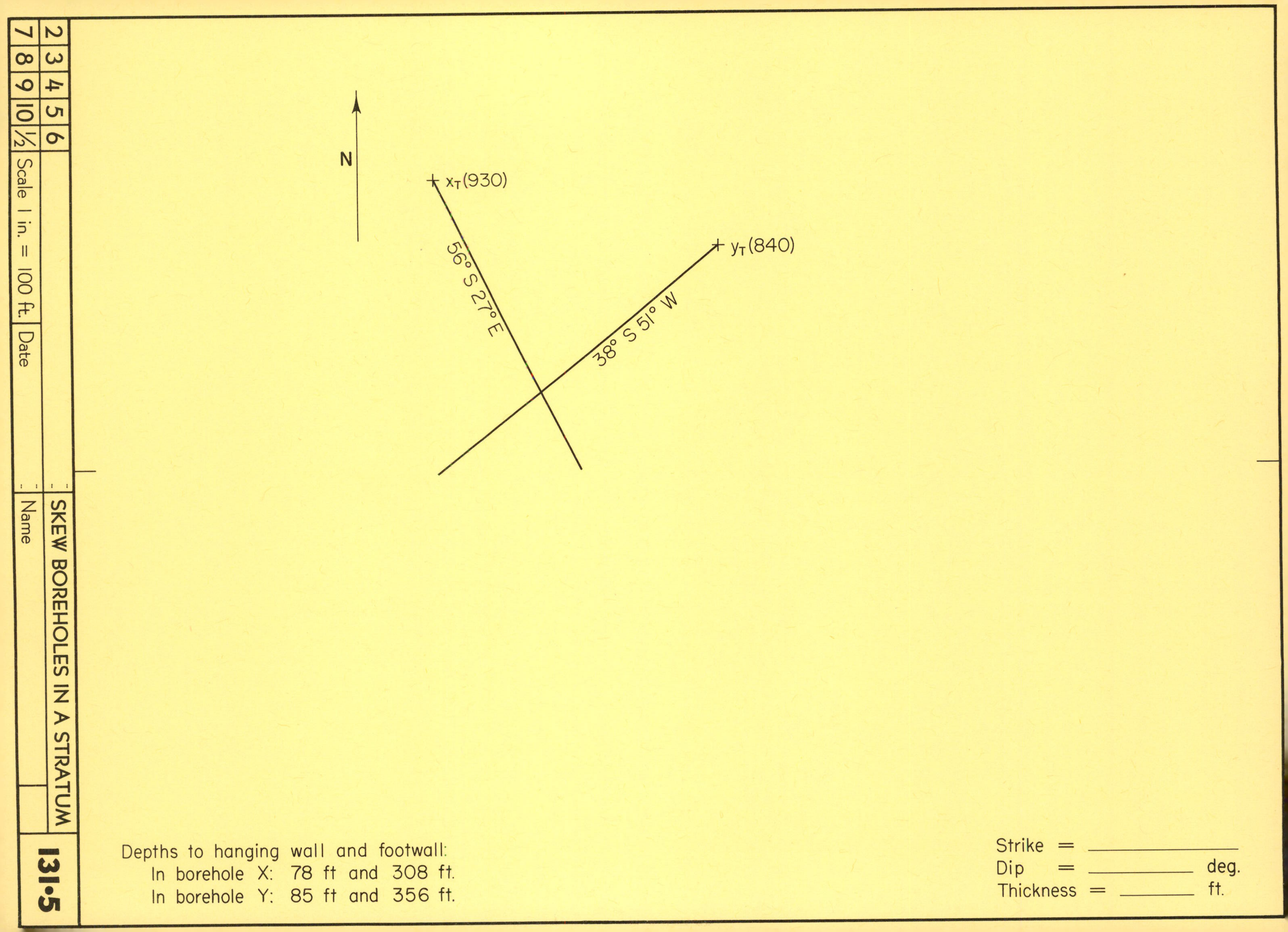
N
x_T(930)
y_T(840)
56° S 27° E
38° S 51° W
Depths to hanging wall and footwall:
In borehole X: 78 ft and 308 ft.
In borehole Y: 85 ft and 356 ft.
Strike = ________
Dip = ________ deg.
Thickness = ________ ft.
2 3 4 5 6
7 8 9 10 ½
Scale 1 in. = 100 ft.
Date
SKEW BOREHOLES IN A STRATUM
Name
131·5

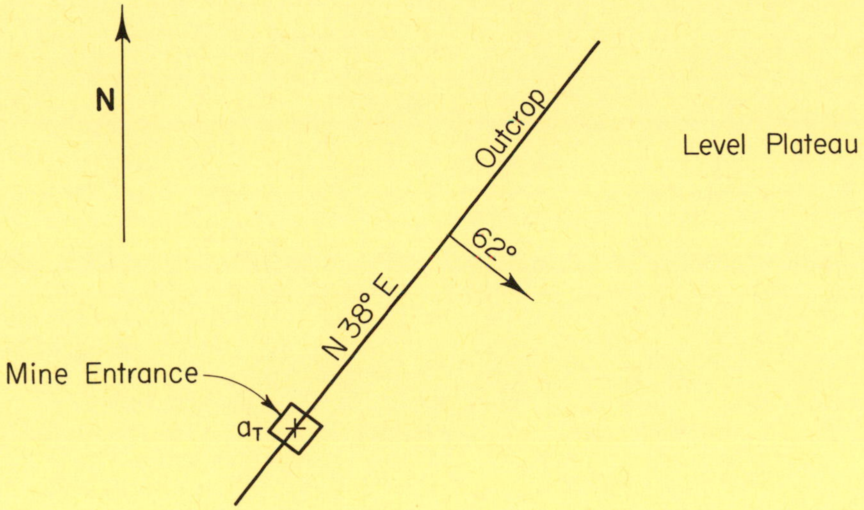

N

Outcrop

Level Plateau

62°

N 38° E

Mine Entrance

a_T

Length of original slope extended to the second vein = ________ ft.

Map distance from A to surface end of the raise = ________ ft. Bearing = ____________

2	3	4	5	6		WORKINGS IN A MINE	132•12
7	8	9	10	½	Scale 1 in. = 50 ft. Date	Name	

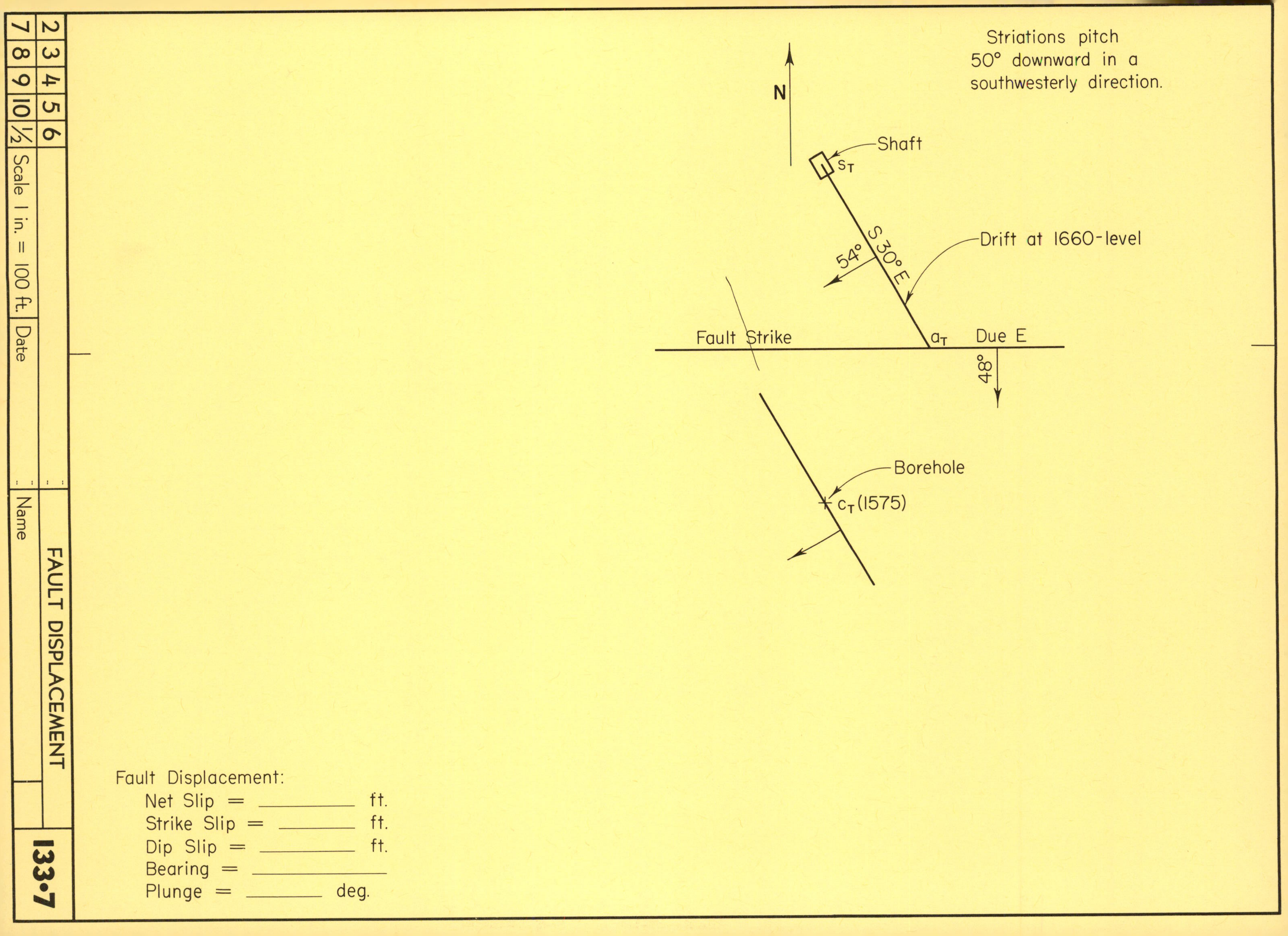

Striations pitch
50° downward in a
southwesterly direction.
N
Shaft
s_T
S 30° E
54°
Drift at 1660-level
Fault Strike
a_T
Due E
48°
Borehole
c_T (1575)
Fault Displacement:
Net Slip = ______ ft.
Strike Slip = ______ ft.
Dip Slip = ______ ft.
Bearing = ______
Plunge = ______ deg.
2 3 4 5 6
7 8 9 10 ½
Scale 1 in. = 100 ft.
Date
Name
FAULT DISPLACEMENT
133•7

A and B are points on upper outcrop line. Test hole at C strikes the upper and lower bedding planes at depths of 20 ft. and 40 ft.

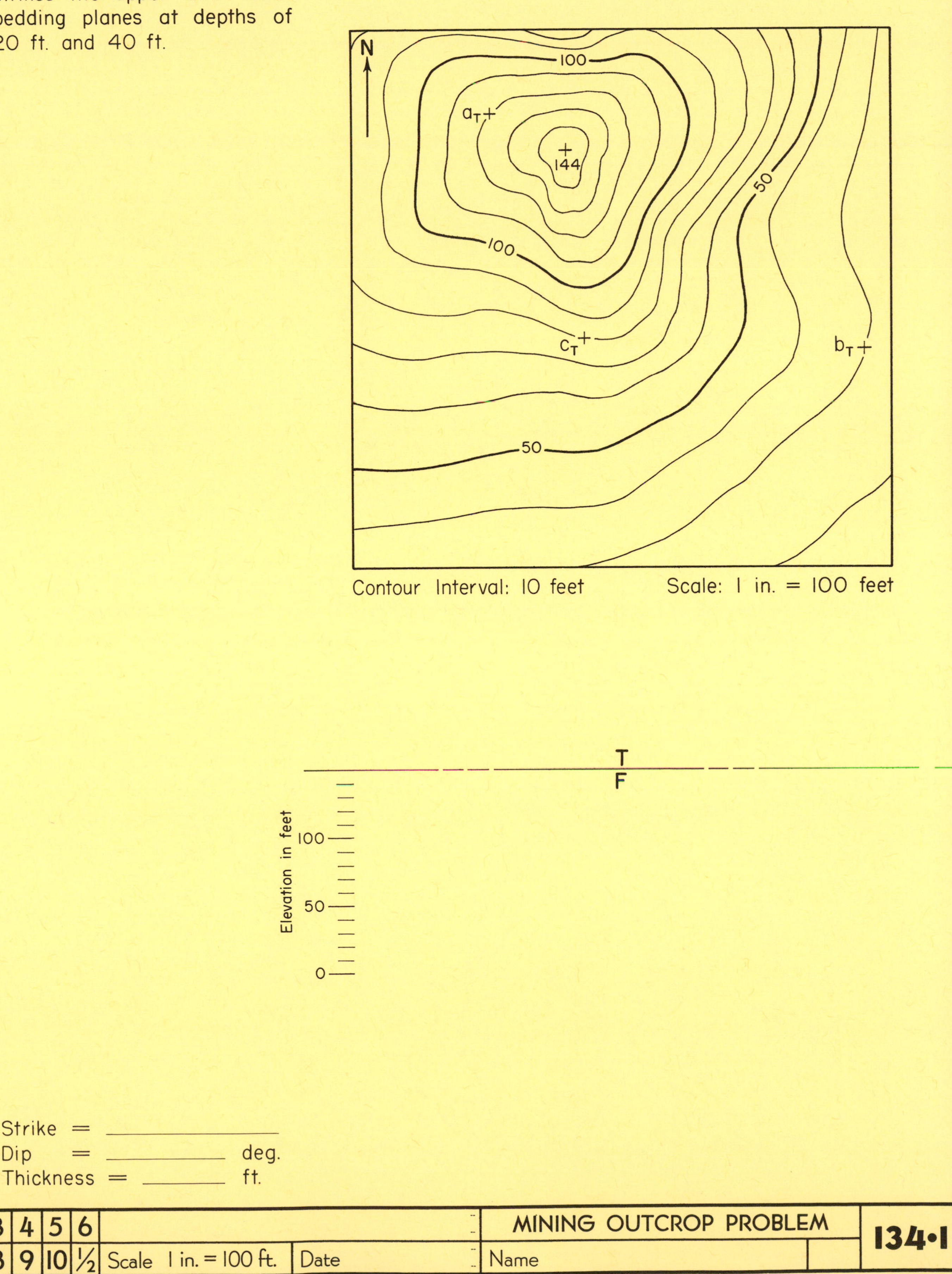

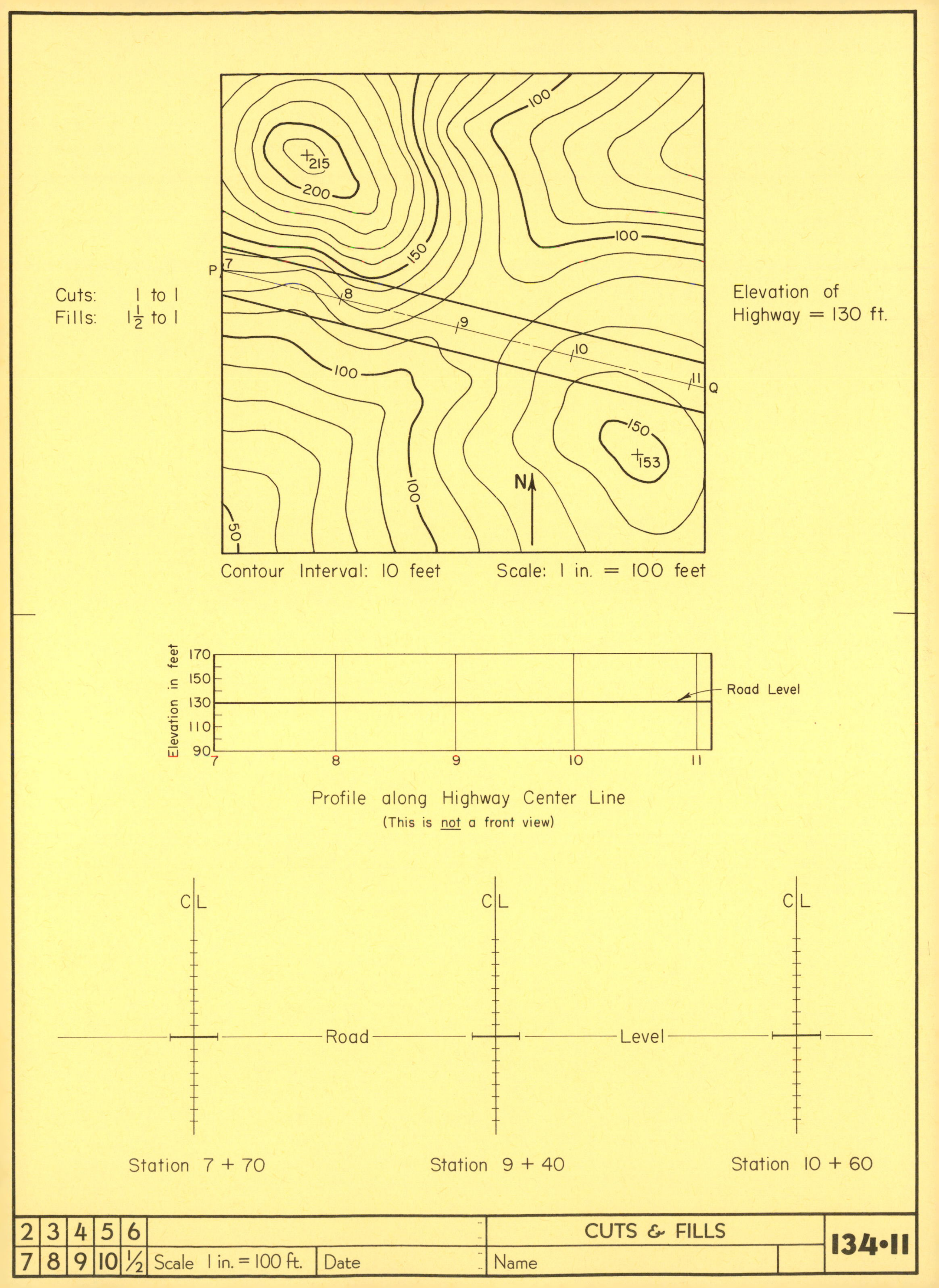

Cuts: 1 to 1
Fills: 1½ to 1
215
200
150
100
100
100
150
153
100
50
P
Q
7
8
9
10
11
N
Elevation of Highway = 130 ft.
Contour Interval: 10 feet
Scale: 1 in. = 100 feet
Elevation in feet
170
150
130
110
90
Road Level
Profile along Highway Center Line
(This is not a front view)
CL
Road
Level
Station 7 + 70
Station 9 + 40
Station 10 + 60
2 3 4 5 6
7 8 9 10 ½
Scale 1 in. = 100 ft.
Date
CUTS & FILLS
Name
134•11

2 3 4 5 6
7 8 9 10 ½ Scale
Date
Name

2	3	4	5	6			
7	8	9	10	½	Scale	Date	Name

2	3	4	5	6			
7	8	9	10	½	Scale	Date	Name

2	3	4	5	6			
7	8	9	10	½	Scale	Date	Name

2	3	4	5	6			
7	8	9	10	½	Scale	Date	Name

2	3	4	5	6			
7	8	9	10	½	Scale	Date	Name

2	3	4	5	6			
7	8	9	10	½	Scale	Date	Name